OPERATIONS MANAGEMENT
With Problems and Solutions.

Rahul Basu

Contents

INTRODUCTION TO OPERATIONS MANAGEMENT 5
PRODUCTION SYSTEM 6
DEFINITION OF OPERATION MANAGEMEN 8
OBJECTIVES OF OPERATION MANAGEMENT 10
FACTORS AFFECTING PRODUCT DESIGN 16
OPERATIONS MANAGEMENT – QUESTION BANK 28
UNIT – I 28
II. OPERATIONS STRATEGY 31
PRODUCTION PLANNING AND CONTROL - CONCEPT 34
CAPACITY PLANNING – CONCEPT 38
CAPACITY REQUIREMENT PLANNING (CRP) 41
FORECASTING – CONCEPT 43
FORECASTING MODELS 45
OPERATIONS MANAGEMENT – QUESTION BANK 51
UNIT – II 51
III PLANT LOCATION & LAYOUT 54
Product Layout (or Line Layout 67
Fixed Position Layout 69
Service Facility Layout 71
LAYOUT DESIGN PROCEDURE 71
CRAFT input requirements are: 72
AUTOMATED LAYOUT DESIGN PROGRAM 73
CORELAP algorithm 75
LINE BALANCING 75
ASSEMBLY LINE BALANCING 77
COMSOAL 77
OPERATIONS MANAGEMENT – QUESTION BANK 78
UNIT – III 78
IV. QUALITY CONTROL 82
Objectives of Quality Control 83
STATISTICAL QUALITY CONTROL 84
CONTROL CHARTS 85
ACCEPTANCE SAMPLING 87
WORK STUDY - CONCEPT 88
Method study 90

WORK MEASUREMENT 93
RELAXATION ALLOWANCE 97
INTERFERENCE ALLOWANCE 97
HUMAN FACTORS IN JOB DESIGN 98
ERGONOMICS 99
QUESTION BANK 104
UNIT – IV 104
CASE STUDY 105
V. MATERIALS MANAGEMENT AND INVENTORY CONTROL 108
MATERIAL REQUIREMENTS PLANNING 111
MRP I vs. MRP II: 114
Meaning of Inventory Control 116
Objectives of Inventory Control 116
ECONOMIC ORDER QUANTITY (EOQ) 118
QUESTION BANK UNIT – V 122
Critical Path Method 129
Pert 138
Calculation of Standard Deviation and Variance for the 150
Critical Activities: 150
NORMAL DISTRIBUTION TABLE 154
Calculation of Total Float for each activity 165
Crashing Of A Project 179
Game Theory, Goal Programming & Queuing Theory 196
Minimax and Maximin Principles 207
Games With No Saddle Point 216
The Principle Of Dominance 226
Method of convex linear combination 231
Graphical Solution Of A 2X2 Game With No Saddle Point 245
2 X N Zero-Sum Games 252
Linear Programming Approach To Game Theory 267
Goal Programming Formulation 284
Queueing Theory 291

INTRODUCTION TO OPERATIONS MANAGEMENT

Introduction and Overview of Operations Management -Understanding similarities and difference among Products, Goods and Services - Historical Development and Current Issues in Operations Management - Operations Strategy and Competitiveness - Product Design – New Product Development, Make or Buy Decisions.

INTRODUCTION

Innovations in technology have resulted in the development of manufacturing capabilities of organisation. Moreover, the study and application of management techniques in managing the affairs of the organisation have also changed its nature over the period of time. Therefore, managing a service system has become a major issue in the global competitive environment. Operations Management has been a driving force in the improvement of business practice around the world. Operations Management leads the way for the organizations to achieve its goals with minimum effort.

Operations management is recognized as an important factor in a country's economic growth. Operation management is the crucial area in the functioning of organizations and therefore, an in-depth study of the subject matter becomes essential. Operation is concerned with the transformation of inputs into the required output or services. Management is the continuous process, which combines and transforms various resources used in the operations system of the organization into value added services. Operation Management is the set of interrelated management activities, which are involved in manufacturing of certain products or services.

Concept of Production:

Production is the step-by-step conversion of one form of material into another form through continuous process to create the utility of the product to the user. Production is a value addition process. Edwood Buffa defines production as a process by which goods and services are created'. Production function is concerned with the transformation of a range of inputs into the required outputs .For example, manufacturing of standardized products like, car, motor cycle, radio, television, soaps, etc.

PRODUCTION SYSTEM

The production system is that part of an organisation, which produces goods of an organisation. It is a planned and integrated activity whereby resources are transformed in a controlled manner to add value for the product. The production system has the following features:

1. Production is a well organized activity with pre-established objectives.
2. The production system converts the various inputs into outputs.

HISTORY OF OPERATIONS MANAGEMENT

The traditional view of manufacturing management began in eighteenth century when Adam smith recognized the economic benefits of specialization of labor. He recommended breaking of jobs down into subtasks and recognizes workers to specialized tasks in which they would become highly skilled and efficient. In the early twentieth century, F.W. Taylor implemented Smith's theories and developed scientific management. From then till 1930, many techniques were developed prevailing the traditional view.

Production Management became the acceptable term from 1930s to 1950s. As F.W. Taylor's works become more widely known, managers developed techniques that focused on economic efficiency in manufacturing. Workers were studied in great detail to eliminate wasteful efforts and achieve greater efficiency.

At the same time, psychologists, socialists and other social scientists began to study people and human behaviour in the working environment. In addition, economists, mathematicians, and computer socialists contributed newer approaches. With the 1970s emerged other two distinct changes.

The most obvious of these, reflected in the new name Operations Management was a shift in the service and manufacturing sectors of the economy. As service sector became more prominent, the change from production 'to operations 'emphasized the broadening of field to service organizations. The second, more suitable change was the beginning of an emphasis on synthesis, rather than just analysis, in management practices. A brief account of development of operations and production management is given below:

Year	Contribution	Contributors
1776	Specialization of labour in manufacturing	Adam Smith
1799	Interchangeable parts, cost accounting	Eli Whitney & others
1832	Division of labour by skill; assignment of jobs by Skill; basics of time study	Charles Babbage
1900	Scientific management time study and work study Developed; dividing planning and doing of work	Frederick W.Taylor
1900	Motion of study of jobs	Frank B. Gilbreth
1901	Scheduling techniques for employees, machines Jobs in manufacturing	Henry L. Gantt
1915	Economic lot sizes for inventory control	F.W. Harris
1927	Human relations; the Hawthorne studies	Elton Mayo
1931	Statistical inference applied to product quality: quality control charts	W.A. Shewart
1935	Statistical Sampling applied to quality control: inspection sampling plans	H.F. Dodge & H.G. Roming
1940	Operations research applications in world war II	P.M. Blacker & others
1946	Digital Computer	John Mauchlly and J.P. Eckert
1950	Mathematical programming, on-linear and stochastic processes	A. Charnes, W.W. Cooper& others
1960	Organisational behaviour: continued study of people at work	L. Cummings, L. Porter
1970	Integrating operations into overall strategy and policy Computer applications to manufacturing, scheduling, and control, Material Requirement Planning (MRP	W. SkinnerJ.Orlicky & G. Wright
1980	Quality and productivity applications from Japan: robotics, CAD-CAM	W.E. Deming & J. Juran

OPERATIONS MANAGEMENT – CONCEPT:

Operation Management is a part of management sciences. Operation Management is concerned with the production of quality goods and services and ensures that the business operations are performed smoothly, efficiently, effectively. It is a field of management that deals with effective planning, scheduling, use and control of a manufacturing or service organisation. Operations management is the business function that plans organizes, coordinates, and controls, the resource needed to produce a company's goods and services. Operations Management is the process whereby resources, flowing within a defined system, are combined and transformed by a controlled manner to add value in accordance with policies communicated by management.

DEFINITION OF OPERATION MANAGEMEN

According to Buffa, production or operation management deals with decision making related to production process so that the resulting goods and services are produced according to specifications, in the amount and by the schedule demanded and at a minimum cost.

The Association of Operation Management defines operation management as, the field of study that focuses on the effective planning ,scheduling, use and control of manufacturing or service organizations through the study of concepts from design engineering, industrial engineering, MIS, quality management, production management, industrial management and other functions as they affect the organisation.

The value addition to an input can be done in the following ways.

They are mentioned below:

OPERATION MANAGEMENT - Core Functions:

Marketing function (communicating)

Product/service development function (creating) Operations function (fulfilling)

Support Functions:

The accounting and finance function.

The human resources function.

Two Meanings of Operations:

- Operations as a function.
- Operations as an activity.

'End-to-end' business processes: Processes that totally fulfill a defined external customer need. **Business process reengineering**: The philosophy that recommends the redesign of processes to fulfill defined external customer needs.

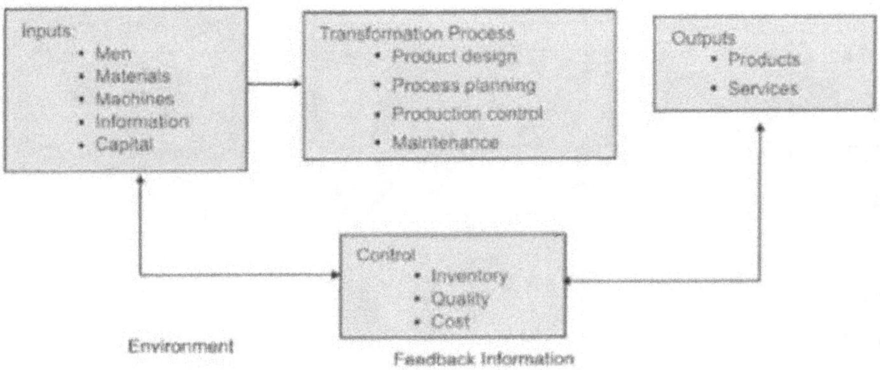

Different characteristics of operations processes:

Volume: repeatability, systemization.

Variety: Standardization, flexible

Variation: variation in demand

Visibility: process exposure

Front office: The high visibility part of an operation

Back office: The low visibility part of an operation

Activities that apply to all types of Operations:

Understanding the operation's strategic objectives

Developing an operations strategy for the

organization Designing an operations strategy for the

organization Planning and controlling the operation

Improving the performance of the operation

The broad responsibilities of operations management

Why is operations management important?

Reduce costs

Increase revenue

Reduce the amount of investment

Provide the basis for future innovation

More profitability

Operations Management:

Operations strategy

Improvement

Planning and control

Design

Operations strategy:

The operation's strategic objectives

The operation's competitive role and position

Improvement

Operations strategy

OBJECTIVES OF OPERATION MANAGEMENT

Operation Management involves management of the entire process responsible for converting inputs into outputs. The following are the objectives of Operations Management.

To provide customer service the main objective of any operating management systems is to utilize resources judiciously for the satisfaction of customer needs and wants. Therefore, customer satisfaction is a key objective of operations management. Operation management focuses on providing the right products at a right price at the right time. Hence, this objective will influence the operations manager's decisions to achieve the required customer service.

Effective utilization of resources that are used in the business organisation must be carefully utilized. Inefficient use of resources or inadequate customer service leads to commercial failure of an organisation. Operations management is concerned essentially with the utilization of resources. It aims at obtaining maximum output from the available resources with minimum cost.

To reduce cost of production Operation management aims at reduction in the cost of production of goods and services. The cost per unit of the product has to be set properly and all efforts should be taken to control the actual cost to pre-determined cost of production. Cost can be classified in to fixed cost and variable cost. The variable cost changes with every level of production. This variable cost can be checked by means of inventory and labor control techniques.

To improve product quality control and maintenance are the two important objectives of operations management. Quality control consists of all those activities, which are designed to define, maintain and control specific quality of products within reasonable limits. It is the systematic regulation of all variables affecting the goodness of the final product. In other words, quality control involves determination of quality standards and its actual measurement .It is necessary to ensure that the established standards are practiced and maintained. It does not attempt to achieve the perfect quality but to secure satisfactory or reasonable quality at a reasonable level of cost.

To fix time schedule Another important objective of operation management is to establish time schedule for various operation activities. The schedule fixation includes the operating cycle time, inventory turnover rate, machine utilization rate, capacity utilization etc.

Proper utilization of Machinery Operation management has to take number of decisions with regard to machinery and equipment. New machines should be installed and the old machines are to be replaced. It has to ensure judicious utilization of machinery and equipment.

Material control based on the sales forecast and production plans, the materials planning and control is done. This involves estimating the individual requirements of parts, preparing materials budget, forecasting the levels of inventories, scheduling the orders and monitoring the performance in relation to production and sales.

What is Role of OM?

OM Transforms inputs to outputs ➢ Inputs are resources such as People, Material, and Money ➢ Outputs are goods and services.

BASIS FOR COMPARISON	GOODS	SERVICES
Meaning	Goods are the material items that can be seen, touched or felt and are ready for sale to the customers.	Services are amenities, facilities, benefits or help provided by other people.
Nature	Tangible	Intangible
Transfer of ownership	Yes	No
Evaluation	Very simple and easy	Complicated
Return	Goods can be returned.	Services cannot be returned back once they are provided.
Separable	Yes, goods can be separated from the seller.	No, services cannot be separated from the service provider.
Variability	Identical	Diversified
Storage	Goods can be stored for use in future or multiple use.	Services cannot be stored.
Production and Consumption	There is a time lag between production and consumption of goods.	Production and Consumption of services occurs simultaneously.

COMPONENTS PRODUCTON FUNCTION

The Product:

It is the most obvious embodiment of the interface between marketing and production and it is not sufficient that the customers require a product; the organization must be capable of producing it. Agreement, therefore, must be reached between all the business functions on matters such as performance, aesthetics, quality, reliability, selling price or production/operating costs, date of delivery and/or time.

The Plant:

A plant of some kind, both in terms of building and equipment is required to manufacture a product. This plant, which accounts for the bulk of the fixed assets of the organization, must continue to be so as long as the consumer's needs can be foreseen. Production management therefore, will be concerned with questions such as future possible demands, design and layout of buildings and offices, performance and reliability of equipment, maintenance of performance, safety of installations and operations, social responsibility, etc.

The Process:

The decision on product or service creation is made by bringing together the technical and organizational needs of the product and the organization and the people within the organization. In deciding upon a process, it is necessary to examine factors such as plant and equipment, safety, maintenance requirements, cost reduction, etc.

The Programme:

A timetable, setting down the date of transfer of products or services to the consumers is the other visible expression of the production/marketing interface, not merely setting down date and time but also determining cash flow, the prime controller of organizational viability. Framing of master timetable further helps to generate timetables for purchasing, maintenance, storage, transport, etc.

The People:

Production from start to finish depends upon people. Like all other products, human beings are variable in intellect, skill and expectations. The work of a social scientist lies in continuously enlarging our understanding of people and organizations and also bringing home better communications, worker participation, industrial democracy, job enrichment, etc. Hence, the production manager should, be involved in discussion on wages/salaries, safety, conditions of work, motivation, training and trade unions.

The Policy:

Production policy is the term applied to those aspects of corporate policy which particularly concern the production or operations departments. Clearly production policy is an integral part of corporate policy and must act within and not independently of it. However, it is the role of the production manager to frame policies, decide on operational details which should be within the organization's capability in determining appropriate production policy.

NEW CONCEPTS AND TRENDS IN OM

Mass Customization

Supply Chain

Management Outsourcing

Lean manufacturing

Agility

Electronic Commerce

PRODUCT DESIGN STAGES:

Every product design passes through the following six stages:

Conception: It is the first stage and the most importance stage in any product design. A draft specification is prepared at this stage by the marketing department consultation with the design department.

Acceptance: Draft specification prepared earlier is scrutinized for its viability by subjecting it to all possible calculations model making, preliminary drawing, laboratory scale processes, etc, if it is not accepted, it may have to be modified or rejected and such decisions are taken jointly by design and marketing department.

Execution: This stage involves the conversion of design specifications into drawings to build the prototype model should be a true replica of the proposed new product satisfying all the requirements of the customer.

Evaluation: The design is evaluated by a cross functional team having representatives from finance, marketing manufacturing and service department to achieve optimal design (the overall best combination of product qualities at the lowest per unit production cost) The design is reviewed to ensure that all requirements of the product such as function, aesthetics (appearance), materials and process alternatives, and their cost, economic assembly , repair and maintenance , lead time required for installing the new process and training the labor etc, are met by the product design.

Translation: In this stage based on the experiences in the previous stages, the detailed engineering drawings for parts, subassemblies, final assemblies, parts lists etc., are prepared. These documents are known as provisional design documents which take into account the Productivity aspects of the design. Also detailed estimates of costs are prepared at this stage.

Preproduction: A pilot production run is carried out using the provisional design documents and the Production aspect in this stage. Based on the experience gained in preproduction, the provisional design is modified in to final design approved for bulk production later.

TYPES OF DESIGN:

Functional design: Functional design involves developing an idea in to a rough model of the proposed products

Aesthetic: Before production on a commercial scale is undertaken another type of design must be integrated with the functional design and its aesthetic design for market acceptability.

Production design or product design: The functional design is translated into production design to make it easy for manufacturing.

Packing design: Packing design should also be appealing to the consumers depending upon the size and nature of the product. Different packing materials can be used to suit the consumers taste and to maintain the chemical properties of the product

CHARACTERISTICS OF A GOOD PRODUCT DESIGN:

A good product design should possess the following characteristics:

Function: A product can be sold if it meets the needs of the consumer and as such the product must be designed to meet such needs. For example, a customer expects a gas lighter to be convenient (i.e., to instantly light the gas stove). If the gas lighter cannot achieve that, then the purpose is lost as the basic function is not met.

Reliability: It is the probability that there will not be any major failure of the product during its use. For example, if certain components of a product are put into use very often, the reliability of each component should be staggeringly high which may not be practical from production point of view. To overcome this problem, duplicate components can be operated in parallel i.e., if one component fails its duplicate may be put into operation.

Maintainability: The lubrication points and other areas for servicing of the product to be designed ought to be easily accessible even though the physical form may have to be altered a bit. The alternative is to make a trouble free product with expensive design. The trade-off between being trouble- free and maintainability is an important decision at the design stage and it mainly depends on the nature of the product. The after sales feedback from customers is quite valuable in improving upon the maintainability of a product.

Producibility: A product should be designed in such a manner that it can be produced easily at a reasonable cost. Least number of operations are required to produce a product quickly and cheaply. This may be possible with change in technology. For example, instead of machining an intricate part such as gear, it can be molded, without affecting its function.

Producibility can also be enhanced by reducing work content of the job involved. For example, the instrument panel of a car used to be maize of wires. This has been redesigned with a printed circuit board where only multiple pin plugs needs to be connected into it. This has resulted into reduced work content and greater product reliability.

Simplification: Simplification and producibility go hand in hand. The simpler the design of the product, the easier it is to produce, the lesser it costs and more reliable it is.

Product standardization and variety reduction: Variety to a large extent depends on market forces. The larger the market, the greater is the degree of standardization possible which makes economies of scale possible. If there is a competitive market, products can have a selling edge if variety is offered. This will involve a design for the product in such a way which will lend itself to modularization so as to gain advantage of large production.

Quality: A good quality product design ensures that the quality of a final product is obtained through its individual components. The tolerance specified at every stage ensures a end product with the desired quality.

Minimum cost: Design influences manufacturing cost. A good product design must ensure minimum manufacturing cost. Some of the areas in which savings can be affected at the design stage.

Warranties: Whether or not a manufacturer is legally bound to put right whatever wrong the customer does is the question here. The point is, a customer is too valuable to be lost, especially if the manufacturer wishes to retain him and established his goodwill. Breakdown of a component may or may not owe their origin to poor maintenance or misuse of the component by a customer it may be due to faulty material, faulty design and faulty processing. The manufacturer must keep in mind the importance of a customer and the costs to be incurred in fixing up a product for the customer.

Modular design: The products should consist of detachable components or sub-assemblies so that whenever a component fails, it can be replaced by a new one easily. The use of standard sub-assembly gives rise to numerous end products in different combinations. The concept is known as modularity. The need of a modular design is enforced by the customer who wants a variety to choose from what he/she likes the best. Maintaining stocks of modules and assembling them promptly enable the manufacturer to offer variety in customer's hands.

Thus, the advantages of modular diversity can be listed as:

Diversity can be offered to the customers Inventories can be substantially reduced. Cannibalization i.e., using parts of one equipment salvaged for use in another application is made easy. Repairing a product becomes in expensive because only a few standard tools are needed. The company can contract or expand its capacity. An example of modular design is a gasoline pump that permits the customer to mix his own blend.

FACTORS AFFECTING PRODUCT DESIGN

1. **Customer requirements:** Product design has to give due respect to the requirements of customers. So designers have to ensure that the product they design suits the convenience of customers for use. The product design has to be in a way that they can be used in all conditions.

2. **Facility to operators:** The designer must see that an operator is provided with all possible comforts and facilities in handling the operations involved in the product design. An operator may not perform efficiently if the product design is so complicated. Operator may get tired off soon and may produce less with a complicated design. Thus machines and tools to perform the operations prescribed in the design must be convenient and comfortable to handle.

3. **Functionality:** A product design should be functionally sound. Purpose of any product design is the complete satisfaction of consumers. So, it has has to function for which it is made. There should be proper coordination in its appearance and service to the consumers.

4. **Cost/price ratio:** Product designers, in a competitive market, face lot of pressures in designing one which is cost effective, better quality and the one that attracts a consumer. Cost effectiveness and quality should be an integral part of a product design. With a constraint on the upper limit on producing products, the designer must ensure cost effective designs.

5. **Product quality:** Quality of a product and its design go hand in hand. Product quality partly depends on how well the design is and partly on conformity. The policy of a firm should stress quality of a product in the design stage itself.

6. **Process capability:** A product design should stress the quality of conformance. The product design should take into consideration the quality of conformance. This depends on how much the equipment and machines can process. So, a product designer should have adequate knowledge of his machinery. He should be capable enough to specify tolerances which can be achieved by the available machines and equipment.

7. **Material requirements:** The nature and quality of the materials have significant impact on the design of the product. The designer should have up to date information about new materials available to make the desired product. Industrial, scientific and technical journals as well as the consumers of the product can provide necessary information in this regard.

8. **Work methods and equipment:** For a most viable design, necessary equipment and method of work carry a significant amount of importance. So, a product designer should be innovative in nature. He should be aware of the innovations for improving the work methods and nature of the equipment.

Production cost can be influenced by a simple modification in work methods. The product design should be such that there is sufficient adaptability and scope for improvement in work methods and type of equipment required for the operations involved in production process.

9. **Effect on existing products:** When a product is designed to replace an existing one, a product designer should consider the use of standard parts and components, manufacturing process, distribution strategies and the techniques to blend new manufacturing technology with the existing product. This will keep the cost of implementing the change and marketing the product to a minimum.

10. **Packaging:** Packaging is an essential part of a product. Design of a package is equally important as the design of the product. Packaging design should consider the objectives of packaging such as protection and promotion of the product. Attractive packaging enhances the sales appeal of product in case of consumer products (non-durables).

NEW PRODUCT DEVELOPMENT – STAGES/ PROCESS:

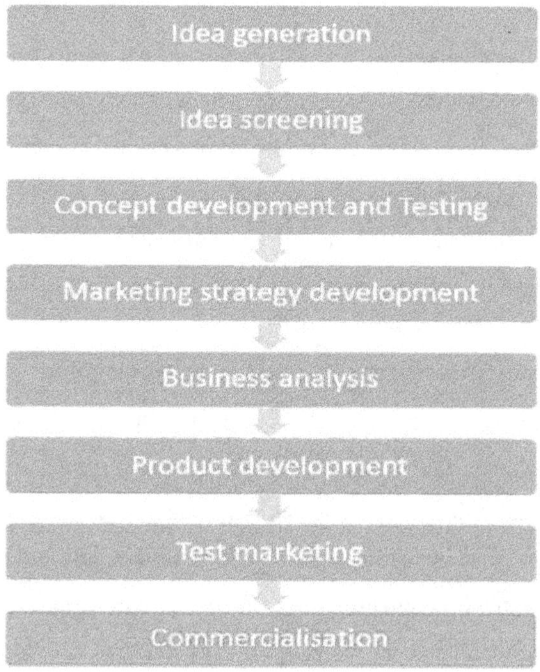

Idea generation:

The new product development process starts with idea generation. Idea generation refers to the systematic search for new-product ideas. Typically, a company generates hundreds of ideas, maybe even thousands, to find a handful of good ones in the end. Two sources of new ideas can be identified: Internal idea sources: the company finds new ideas internally. That means R&D, but also contributions from employees.

External idea sources: the company finds new ideas externally. This refers to all kinds of external sources, e.g. distributors and suppliers, but also competitors. The most important external source are customers, because the new product development process should focus on creating customer value.

Idea screening:

The next step in the new product development process is idea screening. Idea screening means nothing else than filtering the ideas to pick out good ones. In other words, all ideas generated are screened to spot good ones and drop poor ones as soon as possible. While the purpose of idea generation was to create a large number of ideas, the purpose of the succeeding stages is to reduce that number. The reason is that product development costs rise greatly in later stages. Therefore, the company would like to go ahead only with those product ideas that will turn into profitable products. Dropping the poor ideas as soon as possible is, consequently, of crucial importance.

Concept development and Testing:

To go on in the new product development process, attractive ideas must be developed into a product concept. A product concept is a detailed version of the new-product idea stated in meaningful consumer terms. A product concept à a detailed version of the idea stated in meaningful consumer terms A product image à the way consumers perceive an actual or potential product.

Let's investigate the two parts of this stage in more detail.

Concept development

Imagine a car manufacturer that has developed an all-electric car. The idea has passed the idea screening and must now be developed into a concept. The marketer's task is to develop this new product into alternative product concepts. Then, the company can find out how attractive each concept is to customers and choose the best one. Possible product concepts for this electric car could be:

Concept 1: an affordably priced mid-size car designed as a second family car to be used around town for visiting friends and doing shopping.

Concept 2: a mid-priced sporty compact car appealing to young singles and couples.

Concept 3: a high-end midsize utility vehicle appealing to those who like the space SUVs provide but also want an economical car.

As you can see, these concepts need to be quite precise in order to be meaningful. In the next sub-stage, each concept is tested.

Concept testing

New product concepts, such as those given above, need to be tested with groups of target consumers. The concepts can be presented to consumers either symbolically or physically. The question is always: does the particular concept have strong consumer appeal? For some concept tests, a word or picture description might be sufficient. However, to increase the reliability of the test, a more concrete and physical presentation of the product concept may be needed. After exposing the concept to the group of target consumers, they will be asked to answer questions in order to find out the consumer appeal and customer value of each concept.

Marketing strategy development

The next step in the new product development process is the marketing strategy development. When a promising concept has been developed and tested, it is time to design an initial marketing strategy for the new product based on the product concept for introducing this new product to the market. The marketing strategy statement consists of three parts and should be formulated carefully: A description of the target market, the planned value proposition, and the sales, market share and profit goals for the first few years An outline of the product's planned price, distribution and marketing budget for the first year The planned long-term sales, profit goals and the marketing mix strategy.

Business analysis:

Once decided upon a product concept and marketing strategy, management can evaluate the business attractiveness of the proposed new product. The fifth step in the new product development process involves a review of the sales, costs and profit projections for the new product to find out whether these factors satisfy the company's objectives. If they do, the product can be moved on to the product development stage.

In order to estimate sales, the company could look at the sales history of similar products and conduct market surveys. Then, it should be able to estimate minimum and maximum sales to assess the range of risk. When the sales forecast is prepared, the firm can estimate the expected costs and profits for a product, including marketing, R&D, operations etc. All the sales and costs figures together can eventually be used to analyze the new product's financial

attractiveness.

Product development:

The new product development process goes on with the actual product development. Up to this point, for many new product concepts, there may exist only a word description, a drawing or perhaps a rough prototype. But if the product concept passes the business test, it must be developed into a physical product to ensure that the product idea can be turned into a workable market offering. The problem is, though, that at this stage, R&D and engineering costs cause a huge jump in investment. The R&D department will develop and test one or more physical versions of the product concept. Developing a successful prototype, however, can take days, weeks, months or even years, depending on the product and prototype methods.

Also, products often undergo tests to make sure they perform safely and effectively. This can be done by the firm itself or outsourced.

In many cases, marketers involve actual customers in product testing. Consumers can evaluate prototypes and work with pre-release products. Their experiences may be very useful in the product development stage.

Test marketing:

The last stage before commercialization in the new product development process is test marketing. In this stage of the new product development process, the product and its proposed marketing programme are tested in realistic market settings. Therefore, test marketing gives the marketer experience with marketing the product before going to the great expense of full introduction. In fact, it allows the company to test the product and its entire marketing programme, including targeting and positioning strategy, advertising, distributions, packaging etc. before the full investment is made. The amount of test marketing necessary varies with each new product. Especially when introducing a new product requiring a large investment, when the risks are high, or when the firm is not sure of the product or its marketing programme, a lot of test marketing may be carried out.

Commercialization:

Test marketing has given management the information needed to make the final decision: launch or do not launch the new product. The final stage in the new product development process is commercialization. Commercialization means nothing else than introducing a new product into the market. At this point, the highest costs are incurred: the company may need to build or rent a manufacturing facility. Large amounts may be spent on advertising, sales promotion and other marketing efforts in the first year.

MAKE-OR-BUY DECISION

DEFINITION of ' Make-Or-Buy Decision': The act of choosing between manufacturing a product in-house or purchasing it from an external supplier. In a make-or-buy decision, the two most important factors to consider are cost and availability of production capacity.

Factors Influencing Make or Buy Decision:

1. Volume of Production:

The quantity or volume of production affects the make or buy decision to the greater extent. If the volume of production is high, it favors' the make decision and low volume favors buy decisions.

2. Cost Analysis:

The cost analysis refers to the determination of costs to make an item as well as the cost to buy it. The cost to make include – the material cost, direct lab our cost, set up and tooling up costs, depreciation, administrative overheads, interest, insurance, taxes and inventory carrying costs of raw materials and work in process. The cost to make also includes the appropriate allowances, spoilage of work or scrap, and the risk associated with doing business. The cost to buy an item should include -purchase price of the item or component, transportation cost, sales tax and octopi, procurement cost, carrying cost, receiving and incoming inspection costs. The analysis of these two costs helps take decision whether to make or buy.

3. Utilization of Production Capacity:

The organization, which has created large production capacity, favors the decision to make.

4. Integration of Production System:

The vertical integration favors the make decision whereas horizontal integration favors buy decision.

5. Availability of Manpower:

Availability of skilled and competent manpower favors makes decision where as scarce manpower prefers buy decision.

6. Secrecy or Protection of Patent Right:

This condition favors the make decision.

7. Fixed Cost:

A lower fixed cost favors the decision to make and higher fixed cost the make decision.

8. Availability of competent suppliers or vendors

9. Quality and reliability of vendors

FUNCTIONAL ASPECTS OF MAKE OR BUY DECISION:

Make or buy decision should be viewed with both long term and short term perspectives in mind. Some of the effects are tangible and others are intangible. These are classified as follows:

Financial Aspects:

The make decision is always demands an investment in plant, machineries and equipment. The investments can be categorized in to fixed cost and variable cost. The buy decision is associated with only variable cost. Expressing all factors in to money terms carries out a thorough and comparative analysis. Then the decision is to be taken based on which one is more economical, to make or to buy.

Technological Aspects:

The make or buy decision is influenced by:

- ✓ The access to the latest technology to the organization.
- ✓ Feasibility and terms and conditions of technology transfer
- ✓ Outdating of technology
- ✓ Product life cycle

Marketing Aspects:

The marketing aspects have the influence on make or buy decision. When there is a fierce competition, an organization tries to enhance the quality and cut down the costs. The make decision assures the quality and reliability of the parts. Under the situation of increasing market share and a good future sales potential company can have an additional investment potential and hence can opt for make decision. When there is a doubt about the market potential, the company should opt for buy decision.

Purchasing Aspects:

The decision is influenced by:

- ✓ The availability of items or components in sufficient quantities
- ✓ The delivery commitments must be reliably met.
- ✓ The acceptable quality and price level of the product
- ✓ Economy in transportation from the source to the organization
- ✓ Competence and reliability of vendors.
- ✓ Long lasting and mutually rewarding relationships with vendors

Strategic Aspects:

Any decision that is to be taken including make or buy decision should be taken with due consideration to overall objective of the organization. Due importance should be given to the economy, secrecy and flexibility in taking decision regarding make or buy.

EXAMPLE PROBLEMS

1. Demand for the component is at the rate of 6000 per year and this demand is going to continue for next three years. The company has two options. It can get the component manufactured from outside or it can manufacture in house. It costs the company Rs. 2.8 per unit to buy the component. The in house manufacture will incur a fixed cost to the extent of Rs. 10,000 and variable cost of Rs. 1.5 per unit. Give the decision rule for make or buy.

Let x represent the number of units.

The total cost incurred in buying the component is

$$\text{Total cost } (Y) = 2.8\,x$$

Total cost incurred in making the component in house

$$\text{Total cost } (Y) = 10{,}000 + 1.5\,x$$

At BEP, both alternatives result in equal total cost

$\therefore$ At BEP, equation (1) = equation (2)

$$2.8\,x = 10{,}000 + 1.5\,x$$

$$1.3\,x = 10{,}000$$

$$x = 7692.3 \text{ say } 7693$$

The decision rules are:

If the quantity is 7693, both make and buy are equally economical (Results in same costs)

1. Quantity less than 7693, it is economical to buy
2. Quantity more than 7693, it is economical to manufacture

2. The ABC Company is investigating the decision whether to make or buy a plastic packaging, which is currently being purchased at Rs. 7 each.

The demand estimates are shown below:

Demand (Units)	20,000	30,000	40,000	50,000	60,000
Chance (%)	10	30	40	15	5

The decision to manufacture in house costs the company an annual fixed cost of Rs. 80,000 towards renovation and conditioning. Variable costs are estimated at Rs. 5 per unit Give your decision whether to make or buy. At what quantity it is profitable to produce rather than buy.

Solution:

The expected demand (or volume) is determined treating the percentage chance as a probability

Demand (D)	Probability (P)	D.P(D)
20,000	0.10	2,000
30,000	0.30	9,000
40,000	0.40	16,000
50,000	0.15	7,500
60,000	0.05	3,000
	Total	37,500

The Expected Cost to Produce

$$TC = \text{Fixed cost} + \text{Variable cost} \times \text{Quantity}$$
$$= 80{,}000 + 5 \times 37{,}500$$
$$= 2{,}67{,}500$$

The decision is economical to continue to buy.

(*ii*) **Break Even Point**

BEP is the volume of production where the total costs to make is equal to the total costs to buy.

Total cost to make = Total cost to buy

$\therefore\ 80{,}000 + 5x = 7x$

$\therefore\ x = 40{,}000$

It is economical to produce, if the quantity is > 40,000 units.

3. An item, which is required by the company, can be manufactured on any of the three following machines and also it can be purchased at a price of Rs. 1.2 per component. Suggest the best option if the requirement is 2,000 units and give the decision rules.

Solution: Let x be the quantity required.

Machine	Fixed Cost (Rs.)	Variable Cost (Rs/Unit)
M1	[illegible]	0.75
M2	3,500	0.50
M3	92,000	0.06

The cost equations for 4 options are:

The relationship between cost and volume is established. The total costs for all the four alternatives are determined at various quantities (volume).

Expected cost to buy

$$TC = 7 \times 37{,}500 = 2{,}62{,}500$$

The decision is economical to continue to buy.

(*ii*) **Break Even Point**

BEP is the volume of production where the total costs to make is equal to the total costs to buy.

Total cost to make = Total cost to buy

$\therefore\ 80{,}000 + 5x = 7x$

$\therefore\ x = 40{,}000$

It is economical to produce, if the quantity is > 40,000 units.

Alternative	Buy	Machines		
Volume (x)	(TC_B)	TCM_1	TCM_2	TCM_3
10,000	12,000	16,500	40,000	97,000
20,000	24,000	24,000	45,000	93,000
50,000	60,000	46,500	60,000	94,500
60,000	72,000	53,000	65,000	95,000
80,000	96,000	69,000	75,000	96,000
1,00,000	1,20,000	84,000	85,000	97,000
1,10,000	1,32,000	91,000	90,000	97,500
1,20,000	1,44,000	99,000	95,000	98,000
1,30,000	1,56,000	1,06,500	1,00,000	98,500
1,50,000	1,80,000	1,21,000	1,10,000	99,500

The requirement is to the extent of 1, 20,000. The best option is to make on M2, which results in lowest cost.

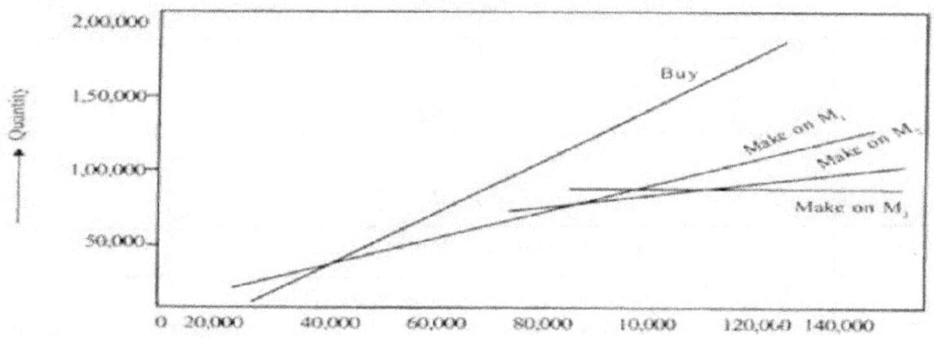

OPERATIONS MANAGEMENT – QUESTION BANK

UNIT – I

PART - A

S. No	Questions (6 marks)	CO	Level
1	Define operations management and state its objectives.	CO1	L1
2	Mention the types of product design.	CO1	L2
3	Recall the characteristics of operations management.	CO1	L1
4	Explain design of production system.	CO1	L1
5	Bring out the core functions of operations management.	CO1	L1
6	Distinguish between goods and services.	CO1	L1
7	Memorize the objectives of designing a product.	CO1	L1
8	State the importance of operations strategy.	CO1	L1
9	State the methods for improving Productivity.	CO1	L2
10	List out the factors influencing make or buy decisions.	CO1	L2

PART – B

S. No	Questions (10 marks)	CO	Level
1	Explain historical approach of operations management.	CO1	L4
2	Evaluate the stages in the development of a new product. Explain each stage in brief?	CO1	L4
3	Explain batch production and mass production along with its advantages and disadvantages.	CO1	L4
4	Highlight the functional aspects of make or buy decisions	CO1	L4
5	Compare and contrast between goods and services.	CO1	L4
6	Discuss the recent trends in production/operations management?	CO1	L5

PART – C

CASE STUDY – (20 marks)

S. No	Questions (20 marks)	CO	Level
1	World's largest car-maker, Toyota Motor (TMC) plans to utilize proposed Indo-Thai Free Trade Agreement (FTA) to make India a hub for small cars to be exported to its global markets. A few auto-parts already enjoy duty free status but the FTA, which is about to be formally inked, will make most auto parts duty free. Toyota imports crucial spares such as the engine for its 'Innova' and 'Corolla Altis' from Thailand. Its top end vehicles Camry Sydan, Prado and Land Cruiser SUV are imported into India. But the company plans to use FTA with Thailand that will allow it to import auto-parts at zero duty when it comes into effect against 7-10% duty it now pays to assemble cars here. TMC says it is looking at all options to make India the manufacturing base for its global operations. Toyota's Indian operations will see a major change once its small cars hit roads in India. Toyota has earmarked Rs. 3,200 crore for year 2008-11 to set up a second plant in Bangalore to make 2 lakh units (cars) from current 80,000 units. Toyota's small cars assembled in India will be exported to overseas. a. Give your analysis about the case. b. Create various ways to solve the problem	CO1	L6
2	A company manufactures and sells gas stoves. It makes some of the parts for the gas stoves and purchases others. The engineering department believes that it is possible to cut costs by manufacturing one of the parts currently being purchased at Rs 8.50 each. The firm uses 1,00,000 of these parts every year and the accounting department compiles the following list of costs based on engineering estimates fixed cost will increase by Rs 50,000,labour cost will rise by Rs 1,25,000.Factory over heads currently running to Rs5, 00,000 per year may increase by 12%. Raw materials used to make the part will cost Rs.6,00,000. Given the above estimate, should the company make the part or continue to buy?	CO1	L6

TEXT / REFERENCE BOOKS

R. Paneerselvam, Production and Operations Management, PHI Learning Private Ltd, 3rd edition, 2012.

R. B. Kanna, Production and Operations Management, PHI Learning Private Ltd, 2nd edition, 2015.

P. Saravanavel & S.Sumathi, Production and Materials Management, Margham Publications, 2nd edition, 2012.

S. N. Chary, Production and Operations Management, Tata McGraw Hill Education Private Limited, 4th edition, 2009.

S. P. Singh, Production and Operations Management, Vikas Publishing House Private Limited, 1st edition, 2014.

II. OPERATIONS STRATEGY

Process Planning – Process Redesigning, Procedure for designing a process - Production Planning and Control– Objectives, Elements, Stages of PPC - Capacity Planning – Importance, Types, Capacity Requirement Planning(CRP) – Forecasting – Factors affecting Forecast, Types of Forecast in Decision Making, Forecasting Models.

PROCESS PLANNING - CONCEPT

A process (manufacturing process) is defined as any group of actions Instrumental to the achievement of the output of an operations system in accordance with a specified measure of effectiveness. When the product is designed, certain specifications are established; physical dimensions, tolerance, standards and quality are set forth. Then it becomes a matter of deciding over the specific details of how to achieve the desired output. This decision is the essence of Process Planning.

The production function essentially is a transformation process that accepts the inputs, and gives the outputs after adding value to the inputs. Process selection is a major strategic decision as it involves allocation of men and material resources as well as financial commitments for a long period. Corporate planning these days, gives enough importance to operations, on par with marketing to which it plays a supportive role. The transformation process is expected to generate an output desired by the market, in the most cost efficient way.

Meaning of Process Planning

Process planning (also known as process designing) is concerned with the specification of process required and its sequence. Process planning is concerned with designing and implementing a work system that will produce the desired product in the required quantities. Process planning consists of all the work that is necessary to arrange for the manufacture of a product by the most economical means and in compliance with all the safety regulations. Process designing begins with receipt of the product specification and ends with the final plans for manufacture of the product. Process planning activities include the type of work flow and the design of work centers.

The objective is to choose and prescribe appropriate methods for manufacturing a product at an optimum cost. Process planning takes into consideration the sale forecasts, order or contracts with customers, drawings and other specifications. The drawings indicate the relationship of parts, models and assembly. Drawings are analyzed to determine the overall scope and sequence of the production process. Process planning is a key decision in production as it affects the efficiency of production. It is essential to design the production process before the actual production takes place. It determines the transformation process after taking into consideration available transformation alternatives involving all possible inputs and outputs. Thus, process planning is the systematic determination of the methods by which a product is to be manufactured, economically and competitively.

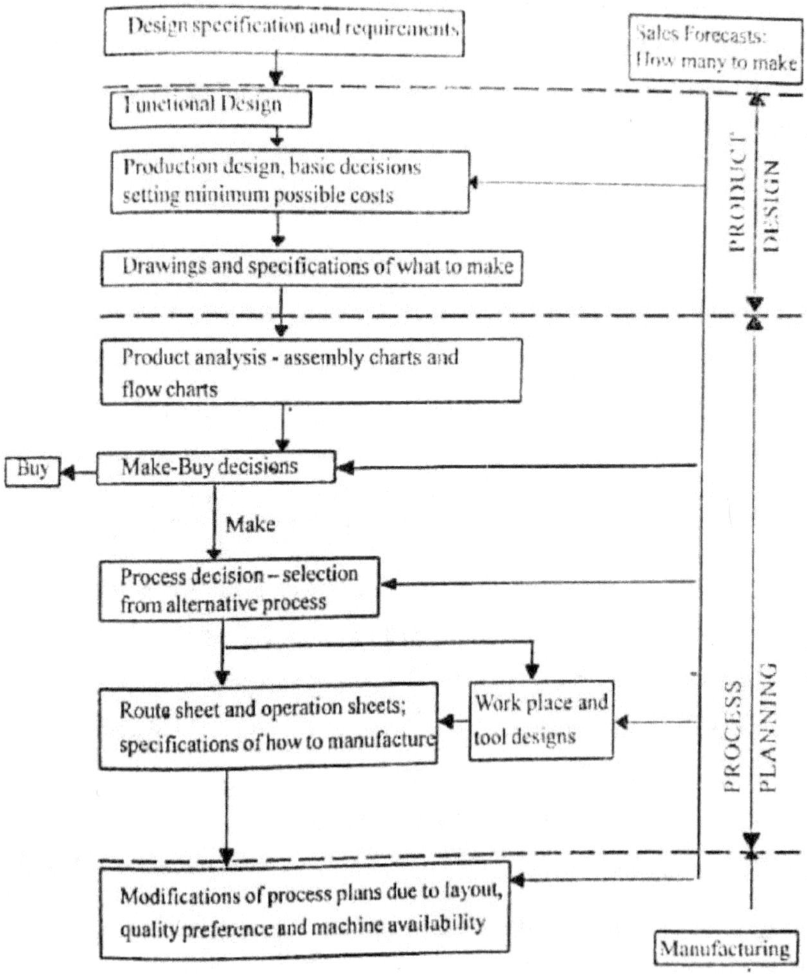

Process Redesigning:

Process designing is not restricted to new concerns or new products only. Any transformation process design is related to a particular situation. With changes in conditions, the transformation process should be redesigned or adapted. While planning a process, industrial engineers have to consider several factors, e.g., efficiency, effectiveness, capacity, lead time, flexibility, etc., these considerations are so interdependent and interwined that a change in one will result in a change in others.

Procedure of Designing a Process:

Product Design: Designing a process begins with the consideration or a careful review of the product design and specifications to ensure that economical manufacturing is feasible. The product engineer and the process planner should collaborate on the design of the product to make sure that realistic specifications are set and the product is designed to permit the use of the most economical method.

Material List: All the materials and parts that will be used are listed. The standard quantity of each item that will be required for manufacturing one unit of final product should be determined. A bill of materials to be bought from outside.

Sequence of Operations: The labor operations to be performed on each component and their sequences (order) are decided. The sequence should be such that it will permit the desired rate and quality of output at the Optimum manufacturing cost. The process engineer must be well experienced in methods and tool design so that the most appropriate sequence of operations can be selected. The method selected must ensure optimization of cost.

Tool Design: The machines, equipment and tools most appropriate for the product and volume of output are then designed. Machine setting e.g., speed, feed temperature, pressure, etc., are also decided at this stage. If a finished product is made up of a large number of components or parts, the sub-assemblies and final assembly are determined to simplify control and to minimize costs.

Layout: The layout of production, installation of manufacturing facilities and auxiliary service is decided. Related operations should be effectively integrated. The grouping of equipment and their proximity to each other would depend upon the volume of production and available production facilities. Close integration of operations makes plant layout compact and reduces time and cost of processing materials. However, it may reduce flexibility of the manufacturing process.

Control System: Necessary control of materials, machines and manpower is established to ensure effective utilization of the manufacturing facilities and most economical production of the product.

The above steps or elements of process designing are usually known as Production engineering or Process engineering or Tool engineering.

PRODUCTION PLANNING AND CONTROL - CONCEPT

Production consists of a sequence of operations that transforms materials from a given form to a desired form (products). The highest efficiency in production is obtained by manufacturing the required quantity of products of the required quality, at the required time, by the best and cheapest method. To achieve this objective, production management employs production planning and control function which is a management tool that coordinates all manufacturing activities. The four factors, Quantity, Quality, Time and Cost encompass the production system of which production planning and control is the nerve center or brain.

There Are Three Stages in PPC

Planning: The choice from several alternatives of the best means of utilizing the resources available to achieve the desired objectives in the most efficient and economic manner.

Action: Performance in accordance with the details set out in the production plan.

Control: The monitoring of performance through a feedback by comparing the results achieved with the planned targets so that performance can be improved through proper corrective action.

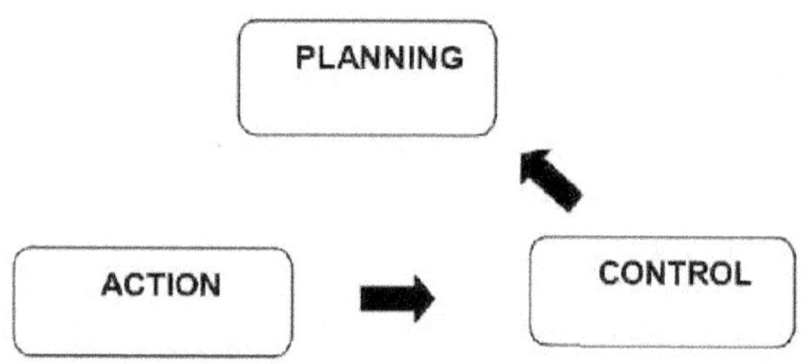

Hence, Production, Planning and Control may be defined as the planning, direction and coordination of the firm's material and physical facilities towards the attainment of predetermined production objectives in the most economical manner. Production, planning and control (PPC) is referred to as operations planning and control because the production planning and control techniques used in production systems manufacturing tangible goods can also be employed in operations or services systems providing services. Production/operation planning and control involves the organization and control of an overall manufacturing system to produce a product (or a service).

Objectives of Production Planning and Control

To ensure maximum utilization of all resources.

To ensure production of quality products.

To minimize the product through put time or production/manufacturing cycle time. To maintain optimum inventory levels.

To maintain flexibility in manufacturing operations.

To co-ordinate between labor machines and various supporting departments.

To plan for capacities between labor and machines and various supporting departments. To plan for plant capacities for future requirements.

To ensure effective cost reduction and cost control.

To prepare production schedules and ensure that promised delivery dates are met. To produce effective results for least total cost.

The ultimate objective is to contribute to profit of the enterprise.

Elements of Production Planning and Control

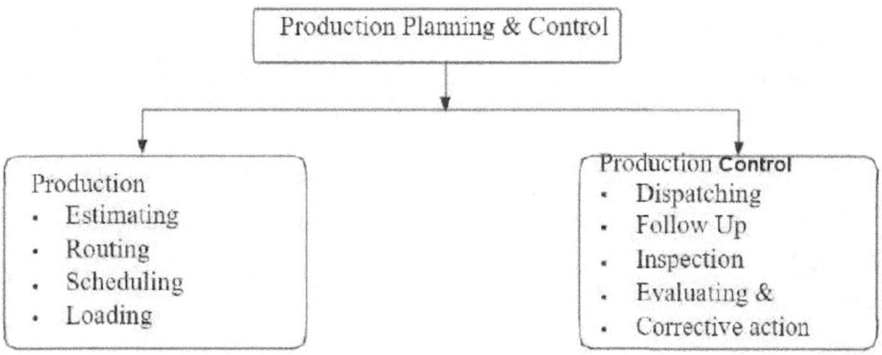

Estimating:

It involves deciding the quality of products to be produced and cost involves in it on the basis of sales forecast.

Estimating manpower, machine capacity and materials required to meet the planned production targets are the key activities before budgeting for resources.

Routing:

This is the process of determining the sequence of operations to be performed in the production process.

Routing determines what work must be done, where and how?

Routing information is provided by product or process engineering function and it is useful to prepare machine loading charts and schedules.

Route sheets: These are the documents providing information and instruction for converting the raw materials into finished products.

Scheduling:

It involves fixing priorities for each job and determining the starting time and finishing time for each operation, the starting dates and finishing dates for each part, sub assembly and final assembly.

Scheduling lays down a time table for production indicating the total time required for the manufacture of a product and also the time required for carrying out the operation for each part on each machine or equipment.

Loading:

Facility loading means loading of facility or work center and deciding which jobs to be assigned to which work center or machine. Loading is the process of converting operation schedules into practice.

Machine loading is the process of assigning specific jobs to machines, men or work centers based in relative priorities and capacity utilization.

A machine loading chart is prepared showing the planned utilization of men and machines by allocating the jobs to machines or workers as per priority sequencing established at the time of scheduling.

Dispatching:

It is defined as, setting production activities in motion through the release of orders and instructions in accordance with the previously planned time schedules and routings.

Dispatching function includes Collecting tools, issuing job orders, obtaining inspection schedules, internal material handling and movement of materials to inspection area after completing the operation, returning jigs and fixtures etc.

Follow up

Expediting or progressing ensures that the work is carried out as per plan and delivery scheduling is met.

Progressing includes activities such as status reporting, attending to bottlenecks or holdups in production and removing the same, controlling variation and deviations from planned performance levels, following up and monitoring progress of work through all stages of production, coordinating with purchase, stores, tool room and maintenance departments and modifying the production plans and re-plan if necessary.

In short, production planning and control function is concerned with decision making regarding:

What to produce: product planning and development includes product design.

How to produce: process planning, material planning, tool planning etc.

Where to produce: facilities planning, capacity planning and sub-contracting planning.

When to produce: production scheduling and machine loading.

Who will produce: man power planning?

How much to produce: Planning for quantity, economic batch size etc.

STAGES IN PPC

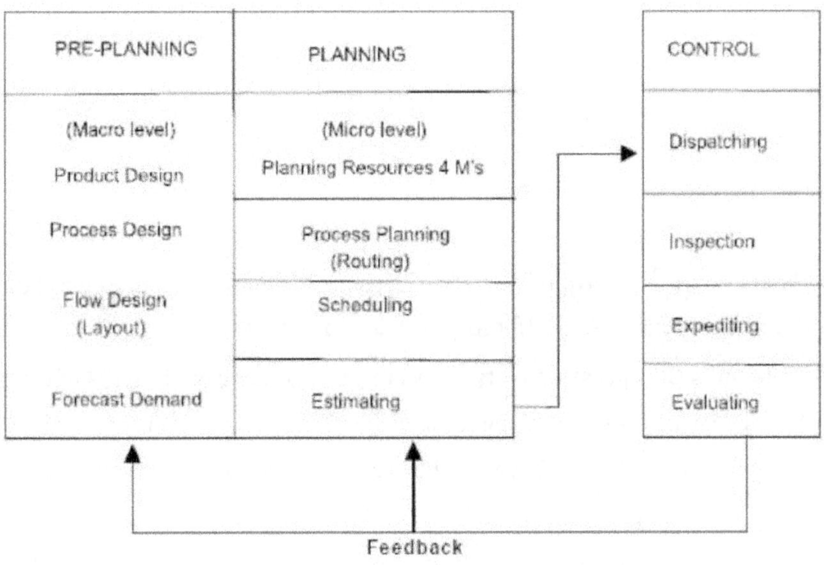

Pre-Planning:

Under this phase of production planning, basic ground work on the product design, layout design and work flow are prepared. The operations relating to the availability scope and capacity of men, money materials, machines, time are estimated.

Planning:

This is a phase where a complete analysis on routing, estimating and scheduling is done. It also tries to find out the areas of concern for short time and long time so that prominent planning can be prepared.

Control:

Under this phase, the functions included are dispatching, follow up, inspection and evaluation. It tries to analyze the expedition of work in progress. This is one of the important phases of the Production Planning and Control.

CAPACITY PLANNING – CONCEPT:

Capacity planning refers to determining what kind of labor and equipment capacities are required and when they are required. Capacity is usually planned on the basis of labor or machine hours available within the plant. Thus, capacity planning is planning for quantity or scale of output. There are four major considerations in capacity planning:

Level of demand

Production

Availability of fund

Management Policy

Importance of Capacity Planning

Capacity limits the rate of output: Therefore, capacity planning determines the ability of an enterprise to meet future demand for its products and services.

Capacity influences the operating costs: Capacity is determined on the basis of estimated demand. Actual demand is often different from estimated demand. As a result, there arises excess capacity or under capacity. Excess or idle capacity increases the cost per unit of output. Whereas under capacity results in the loss of sales.

Capacity decisions leave a direct impact on the amount of fixed investment made initially.

Capacity decisions result in Long-term commitment of funds. Such long-term decisions cannot be reversed except at major costs.

Types of Capacity

Licensed capacity - denotes the capacity licensed by the Government authorities concerned.

Potential capacity - The decision on potential capacity is taken mostly by a senior most executive of the organization.

Immediate capacity - is that which can be made available within the current budget period. Immediate capacity is subject to certain constrains like plant equipment size, availability of equipment, availability of manpower, financial policy, sub-contracting policy, the technical demands of the tasks, and the number of different tasks being undertaken. For example, the capacity of a restaurant is limited by the size of the dining area or the number of tables.

Design or Installed capacity - It is the maximum output that can be achieved in a given time period form a particular plant. It is a theoretical capacity as it does not take into consideration power breakdown, pool planning, non-availability of materials, labor absenteeism, etc. This capacity is reliable only if certain conditions are satisfied. They are:

- There are no interruptions of any kind.
- There is 100% utilization of capacity.
- Men and machines work in ideal conditions.
- Quality of inputs according to specification.

In real life situation, it is difficult to fulfill these conditions. Therefore, installed capacity only sets the maximum limit and also serves to judge the actual utilization of plant capacity. Hence it is also known as maximum capacity.

Effective or Practical or Operating capacity: Effective capacity can be influenced by technical abilities in the pre-operations stages, organizational skills in the planning stages, purchasing skills, sub-contracting skills, maintenance policies and abilities, efficiency of workforce, multiple shift operation, etc. No plant can work up to the maximum or theoretical capacity due to plant efficiency factor and scrap factor. A portion of the available hours cannot be worked due to scheduling delays, machine breakdown, preventive maintenance, etc. This results in the efficiency of plant being less than what is rated.

Actual or Utilized capacity: This is the actual output achieved during a particular time period.

If installed capacity is 100,000 tonnes and the actual production is 80,000 tonnes, we say that capacity utilization is 80% or the plant worked at 80% of the capacity.

Normal Capacity or Rated capacity: It is the capacity estimated by a qualified authority as to the amount of production that should be usually secured Actual capacity is usually expressed as a percentage of the rated capacity.

Excess capacity: Generally plant and equipment are indivisible in nature. Plant and equipment are long term facilities and constitute the major part of production cost. It is not possible to adjust fully and immediately the size of the plant and machinery to suit day to day changes in sales and production. Therefore, excess (unutilized) capacity may occur frequently.

Procedure for Capacity Planning

Assessment of Existing Capacity: Capacity of a unit can be measured in terms of output or inputs. Output measure is appropriate in case of manufacturing concerns. e.g., automobile plant (number of cars), iron and steel plant (tons of steel), cannery (tons of food), etc. Service concerns like hospitals (number of beds), theatres (number of seats), etc., can measure capacity in terms of inputs.

Forecasting Future Capacity Needs: Short term capacity requirements can be estimated by forecasting product demand at different stages of the product life cycle. It is more difficult to anticipate long-term capacity requirements due to uncertainties of market and technology. Capacity forecast helps to determine the gap between the existing capacity and estimated capacity so that necessary adjustments may be made.

Identifying Alternative ways of Modifying Capacity: In case where the existing capacity is inadequate to meet the forecast demand capacity, the expansion is required to meet the shortage. Additional shifts may be employed to expand the capacity. Expansion will provide economies of scale and help in meeting the forecast demand. But it involves additional investment and danger of fall in forecast demand in future. When the existing capacity exceeds forecast capacity, there is a need for reduction of excess capacity. Developing new products, selling of existing facilities, layout of workers or getting work from other firms are the methods of overcoming it.

Evaluation of Alternatives: Various alternatives for capacity expansion or reduction are evaluated from economic, technical and other viewpoints. Reactions of employees and local community should also be considered. Cost-Benefit analysis, Decision theory and Queuing theory are the main techniques of evaluating alternatives.

Choice of Suitable Course of Action: After performing the cost-benefit analysis of various alternatives to expand or reduce the capacity, the most appropriate alternative is selected.

CAPACITY REQUIREMENT PLANNING (CRP) – CONCEPT:

Capacity is a measure of the productive capability of a facility per unit of time. Capacity decisions begin with the initial facility layout and extend to aggregate planning, master scheduling, capacity requirements planning .CRP is a technique for determining what personnel and equipment capacities are needed to meet the production objectives embodied in the master Schedule and the material requirements plan.

CRP is an effort to develop a match between the MRP schedule and the production capacity of the company. Determination of the capacity of the work center and the capacity requirements imposed on those work centers by a particular product mix enables a company to know the level of sales its production system can support. Thus, company will be able to make realistic sales commitments. Capacity planning helps to avoid under-utilization of capacity and also CRP enables the company to anticipate production bottlenecks in some work centers in time to take corrective actions.

To be effective, capacity requirements planning must be coordinated with MRP. Working together, MRP and CRP programmes translate the master schedule to requirements for components and capacity, simulating the impact of the master schedule that provided the input for MRP programme. CRP can be used to refine the master production schedule (MPS) further after MRP is run.

Inputs for CRP Process:

Planned orders and released orders from the MRP system.

Loading information from work center status file.

Routing information from the shop routing file.

Changes which modify capacity give alternative routings or alter planned orders.

The released and planned orders from the MRP system are converted into standard hours of load by the CRP system. MRP system assumed that capacity is available when needed unless otherwise indicated.

Outputs of CRP:

Apart from information for modification of capacity or revision of MPS, the major outputs of the CRP system are the verification of planned orders for the MRP system and load reports. The firm can plan for the average amount of labor and equipment that is expected without actually designating the capacity for specific orders. The flow of information in CRP is shown;

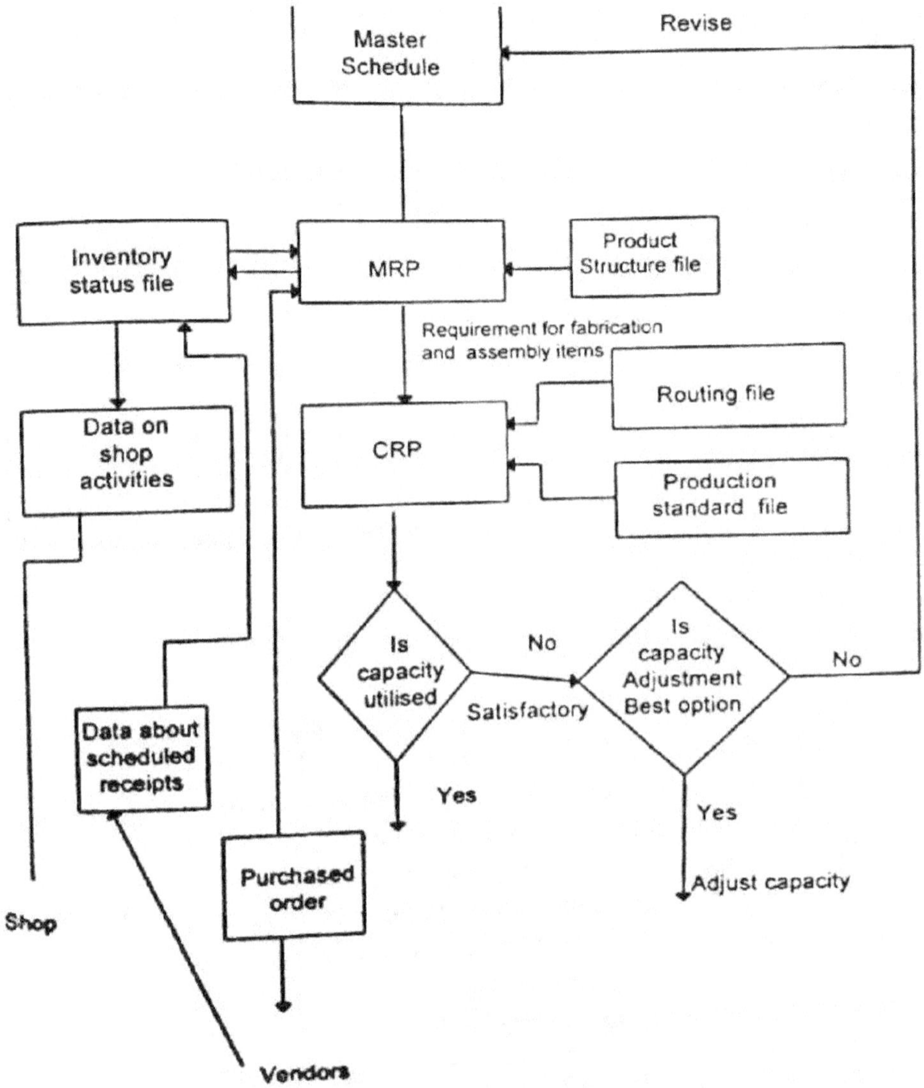

FORECASTING – CONCEPT:

A forecast is an estimate of an event which will happen in future. The event may be demand of a product, rainfall at a particular place, population of a country, or growth of a technology. The forecast value is not a deterministic quantity. Since, it is only an estimate based on the past data related to a particular event, proper care must be given in estimating it.

In any industrial enterprise, forecasting is the first level decision activity. That is the demand of a particular product must be available before taking up any other decision problems like, materials planning, scheduling, type of production system (Mass or batch production) to be implemented, etc.

So, forecasting provides a basis for coordination of plans for activities in various parts of a company. All the functional managers in any organization will base their decisions on the forecast value. So, it is vital information for the organization. Due to these reasons, proper care should be exercised while estimating forecast values.

In business, forecasts may be classified into technology forecasts, economic forecasts and demand forecasts.

Technology Forecast:

Technology is a combination of hardware and software. Hardware is any physical product while software is the know-how, technique or procedure. Technology forecast deals with certain characteristics such as level of technical performance, rate of technological advances. Technological forecast is a prediction of the future characteristics of useful machines, products, process, procedures or techniques. Based on the importance of this activity, Government of India has established a "Technology Information Forecasting and Assessment Council (TIFAC)", under the Ministry of Science and Technology to promote action oriented studies and forecasting in selected areas.

Economic Forecasts:

Government agencies and other organizations involve in collecting data and prediction of estimate on the general business environment. These will be useful to government agencies in predicting future tax revenues, level of business growth, level of employment, level of inflation, etc. Also, these will be useful to business circles to plan their future activities based on the level of business growth.

Demand Forecast:

The demand forecast gives the expected level of demand for goods or services. This is the basic input for business planning and control. Hence, the decisions for all the functions of any corporate house are influenced by the demand forecast.

Factors Affecting Forecast (Demand):

Business cycle
Random variation
Customer's plan
Product's life cycle
Competition's efforts and prices
Customer's confidence and attitude
Quality
Credit policy
Design of goods or services
Reputation for service
Sales effort
Advertising

Types of Forecasting In Decision Making

Forecasting in different functional areas of management such as Marketing, Production, Finance and Personnel play a crucial role for planning ahead. The various types of forecast in each of the aforementioned areas are as follows:

1. **Marketing**

Demand forecasting of products

Forecast of market share

Forecasting trend in prices

2. **Production**

Forecast of Materials requirements

Forecast of Trends in material and labor

costs Forecast of Maintenance requirements

Forecast of Plant capacity

3. **Finance:**

Forecast of Cash flows

Forecast of Rates of expenses

Forecast of Revenues

4. Personnel:

Forecast of Number of workers in each

category Forecast of Labor turn over

Forecast of Absenteeism

FORECASTING MODELS

The forecasting techniques can be classified into qualitative techniques and quantitative techniques. These are presented below. Qualitative techniques use subjective approaches. These are useful where no data is available and are useful for new products. Quantitative techniques are based on historical data. These are more accurate and computers can be used to speed up the process.

Quantitative Forecasting Techniques

Simple moving average
Single exponential smoothing
Double moving average
Double exponential smoothing
Simple regression
Semi-average method
Multiple regression
Box Jenkins

Qualitative Forecasting Techniques

Delphi type method
Market surveys

Selection of a Forecasting Technique:

The selection of a forecasting technique depends on the following three factors:

I. The characteristics of the decision making situation, which include:
The time horizon
Level of detail
Number of items
Control versus Planning

II. The characteristics of the forecasting methods:
The time horizon (number of periods for which forecasting
required) The pattern of data (horizontal, seasonal, trend, etc.)

Type of model (casual, time series or statistical) Cost

Accuracy

Ease of application

III. Present situation, which includes:

The item that is being forecast

Amount of historical data available

Time allowed for preparing forecast

Measures of Forecast Accuracy:

Demand forecast influences most of the decisions in all the functions. Hence, it must be estimated with the highest level of precision. Some common measures are inevitable to measure the accuracy of a forecasting technique. This measure may be an aggregate error (deviation) of the forecast values from the actual demands. The different types of errors which are generally computed are as presented below.

Mean Absolute Deviation (MAD):

It is the mean of absolute deviations of forecast demands from actual demand values. The MAD is sometimes called as the mean absolute error (MAE).

Mean Square Error (MSE):

Mean square error is the mean of the squares of the deviations of the forecast demands from the actual demand values. Usually the effects on operations of small errors are not serious. These errors may be smoothed out by inventory or overtime work. It will be difficult to have smoothed values for forecast even if there are few large errors. Consequently, a method of measuring errors that penalizes large errors more than small errors is sometime desired. The mean square error (MSE) provides this type of measure of forecast error.

Mean Forecast Error (MFE):

Mean forecast error (MFE) is the mean of the deviations of the forecast demands from the actual demands.

Mean Absolute Percentage Error (MAPE):

Mean absolute percentage error (MAPE) is the mean of the percent deviations of the forecast demands from the actual demands.

Simple Moving Average Method:

A simple moving average is a method of computing the average of a specified number of the most recent data values in a series.

Weighted Moving Average:

Equal weights were assigned to all periods in the computation of the simple moving average. The weighted moving average assigns more weight to some demand values (usually the more recent ones) than to others.

Simple (Single) Exponential Smoothing:

Another form of weighted moving average is the exponential smoothed average. This method keeps a running average of demand and adjusts it for each period in proportion to the difference between the latest actual demand figure and the latest value of the average.

Adjusted Exponential Smoothing:

The simple exponential smoothing forecast is a smoothed average positioned on the current period. It is taken as a next period forecast. In reality, trend exists in demand pattern of much business. Hence, due recognition should be given to make correction in the demand forecast for trend also. Adjusted exponential smoothed forecast model actually projects the next project forecast by adding a trend component to the current period smoothed forecast.

Linear Regression:

Regression means dependence and involves estimating the value of a dependent variable Y, from an independent variable X. In simple regression, only one independent variable is used, whereas in multiple regression on two or more independent variables are involved.

Semi-average Method:

This method is sometimes employed when a line appears to be an inadequate explanation of the trend. According to this method, the original data are divided into two equal parts and the values of each part are then summed up and averaged. The average of each part is centered in the period of the time of the part from which it has been calculated and then plotted on a graph. Then a straight line is drawn to pass through the plotted points. This line constitutes the semi- average trend line. When the number of years is odd, the middle year is not considered while dividing the data into two equal parts and obtaining the average.

Delphi Method:

Delphi method is a forecasting technique applied to subjective nature of demand values. In view of globalization in India, Indian companies will have difficulty in estimating the demand of their products mainly because of possible mixed reactions of customers towards various attributes of a specific product which is manufactured by multinational firms and indigenous firms. Under such situation, one has to resort to subjective estimates. Technology forecasting is another example where there is no quantitative data based on which the future technology can be predicted. In this situation, we will have information at various stages of technological advancement for a particular application. If we closely examine the development of computer languages, the following is the order of development.

- Machine Language.

- Assembly Language (First Generation Languages)
- High Level Languages (Second Generation Languages)
- Third Generation Languages (Dbase-III, Lotus 1-2-3)
- Fourth Generation Languages (Oracle, Sybase, PC-Focus)

Beyond this stage, the next level development would be clubbed under Fifth Generation languages. But the features of such languages are yet to be known. If the objective is to predict the capabilities of such languages, one has to use Delphi method.

In Delphi method of forecasting, several knowledgeable persons are asked to provide subjective estimates of demands or forecasts of possible advances of technology. The experts may provide several opinions. Based on the Opinions of the experts, a consensus will be arrived at the demand of product/advances of technology.

The essential precautions to be followed in this method are as follows:

- Panel members must be unknown to each other.
- The initial questionnaire should be unambiguous and it should explain every matter about which opinion is sought.

After getting the opinions from the panel members, they are to be compared for similarity. If the variation among the opinions is too much, the summary of opinions is to be circulated again among the members without mentioning the names of persons who provided the opinions. Generally, 50% of the estimate is treated as the basis for comparison. The panel members whose opinions differ significantly from the middle 50% of the estimate will be asked to reconsider their opinions. Still, if they want to stick to their original opinions, they will be asked to provide rationale for the same.

So, the Delphi method is an iterative process until the panel converges on a specific value or a range of values as defined by the required accuracy, or arrives at a consensus on the matter under consideration.

—

OPERATIONS MANAGEMENT – QUESTION BANK

UNIT – II

PART – A

S. No	Questions (5 marks)	CO	Level
1	Discuss the factors influencing product design.	CO2	L1
2	Give reasons for the process redesigning.	CO2	L2
3	Mention the three stages of PPC.	CO2	L3
4	Explain the concept of CRP.	CO2	L2
5	List down the objectives of PPC.	CO2	L2
6	State the meaning of process planning.	CO2	L3
7	Memorize the importance of capacity planning.	CO2	L3
8	What are the factors affecting forecasting demand?	CO2	L3
9	Bring out the types of forecasting areas in decision making.	CO2	L4
10	List any four disadvantages of market research method of demand forecasting.	CO2	L3

PART – B

S. NO	Questions (10 marks)	CO	Level
1	Elaborate the procedure of designing a process.	CO2	L4
2	Enumerate the stages in PPC.	CO2	L4
3	Draw diagram and explain capacity requirement planning.	CO2	L4
4	Briefly explain the forecasting methods based on business context.	CO2	L5
5	Elaborate the quantitative forecasting techniques.	CO2	L6
6	Analyze the qualitative forecasting technique.	CO2	L6

PART – C

CASE STUDY – (20 marks)

S. No	Questions (20 marks)	CO	Level
1	Mahindra and Mahindra hired, Prof. Yasutoshi Washio, a Japanese expert, to implement Deming guidelines in the company. Prof. Yasutoshi Washio was skeptical about the Indian companies and workers. He felt that the Indian companies are more like the American companies, which feel that results are important. On the other hand, for the Japanese, the process is more important. Moreover, he had serious doubts about the attitude of the Indians workers with respect to a team work - a Deming prerequisite - as he felt that Indians were individualistic. In the initial few years of interaction with the management of FES, Washio found himself isolated due to disagreements on various fronts. Washio has major difficulties in making most of the Indian companies understand the importance of implementation over creating a perfect framework a. Give your analysis about the case, b. Create various process planning techniques.	CO2	L6
2	The GM (Works) has problems with manufacturing budgets, meeting cost reduction targets, and dealing with new products manufacturing schedules. When an in-depth interview (non-directive type) was conducted between the GM (Works) and the Chairman of the Company, the GM (Works) explained that many things are happening in the Company about which he is ignorant, particularly the preparation, new product integration, etc. He agrees to the view that the Company is interested in high-growth and high-profit, but he has never been given an opportunity to review his own scheme of things and explain to the top management. The production culture of the company has never been assessed whereas the stringent rules are being directed by the finance and personnel departments. And sometimes, show cause notices are being served to supervisors and	CO2	L6

	senior employees. The Company is introducing new products without assessing the capability of the manufacturing system and the resources. (a) Under the above situation, if you are asked to work as a consultant to show the perspectives to the Board of Management, what action plans would you suggest? (b) Does Business Process Re-engineering (BPR) help in situations like these?		

TEXT / REFERENCE BOOKS

R. Paneerselvam, Production and Operations Management, PHI Learning Private Ltd, 3rd edition, 2012.

R. B. Kanna, Production and Operations Management, PHI Learning Private Ltd, 2nd edition, 2015.

P. Saravanavel & S.Sumathi, Production and Materials Management, Margham Publications, 2nd edition, 2012.

S. N. Chary, Production and Operations Management, Tata McGraw Hill Education Private Limited, 4th edition, 2009.

S. P. Singh, Production and Operations Management, Vikas Publishing House Private Limited, 1st edition, 2014.

III PLANT LOCATION & LAYOUT

Facility Location – Factors influencing Plant Location, Break even Analysis. Plant Layout – Classification of Layout, Layout Design Procedures – CRAFT, ALDEP, CORELAP. Line Balancing – Objectives of Assembly Line Balancing, Ranked Positional Weight Method, COMSOAL

PLANT LOCATION – CONCEPT

Plant location is defined as deciding a suitable location or place etc., where the factory or plant will start functioning. Plant location or the facilities location problem is an important strategic level decision- making for an organisation. One of the key features of a conversion process (manufacturing system) is the efficiency with which the products (services) are transferred to the customers. This fact will include the determination of where to place the plant or facility.

The selection of location is a key-decision as large investment is made in building plant and machinery. It is not advisable or not possible to change the location very often. So an improper location of plant may lead to waste of all the investments made in building and machinery, equipment.

Before a location for a plant is selected, long range forecasts should be made anticipating future needs of the company. The plant location should be based on the company's expansion plan and policy, diversification plan for the products, changing market conditions, the changing sources of raw materials and many other factors that influence the choice of the location decision. The purpose of the location study is to find an optimum location one that will result in the greatest advantage to the organization.

Objectives of plant location:

Reduced capital investment and operating

cost Ensuring effective plant layout

Coordination with government policies

Employee welfare and public needs Security

Aspects of plant location:

There are three aspects;

Selection of a region

Selection of a locality (community factor)

Selection of a site

Need for location selection:

I. When starting a new organization, i.e., location choice for the first time.

II. In case of existing organization.

III. In case of Global Location.

I. In case of location choice for the first time or new organizations:

Cost economies are always important while selecting a location for the first time, but should keep in mind the cost of long-term business/organisational objectives. The following are the factors to be considered while selecting the location for the new organizations:

Identification of region: The organisational objectives along with the various long-term considerations about marketing, technology, internal organisational strengths and weaknesses, region specific resources and business environment, legal-governmental environment, social environment and geographical environment suggest a suitable region for locating the operations facility.

Choice of a site within a region: Once the suitable region is identified, the next step is choosing the best site from an available set. Choice of a site is less dependent on the organisation's long-term strategies. Evaluation of alternative sites for their tangible and intangible costs will resolve facilities-location problem. The problem of location of a site within the region can be approached with the following cost-oriented non-interactive model, *i.e.*, dimensional analysis.

II. In case of location choice for existing organization:

In this case a manufacturing plant has to fit into a multi-plant operations strategy. That is, additional plant location in the same premises and elsewhere under following circumstances:

Plant manufacturing distinct products.

Manufacturing plant supplying to specific market area.

Plant divided on the basis of the process or stages in

manufacturing. Plants emphasizing flexibility.

The different operations strategies under the above circumstances could be:

Plants manufacturing distinct products: Each plant services the entire market area for the organization. This strategy is necessary where the needs of technological and resource inputs are specialized or distinctively different for the different product-lines.

For example, a high quality precision product-line should not be located along with other product-line requiring little emphasis on precision. It may not be proper to have too many contradictions such as sophisticated and old equipment, highly skilled and semi-skilled personnel, delicate processes and those that could permit rough handlings, all under one roof and one set of managers. Such a setting leads to much confusion regarding the required emphasis and the management policies.

Product specialization may be necessary in a highly competitive market. It may be necessary to exploit the special resources of a particular geographical area. The more decentralized these pairs are in terms of the management and in terms of their physical location, the better would be the planning and control and the utilization of the resources.

Manufacturing plants supplying to a specific market area: Here, each plant manufactures almost all of the company's products. This type of strategy is useful where market proximity consideration dominates the resources and technology considerations. This strategy requires great deal of coordination from the corporate office. An extreme example of this strategy is that of soft drinks bottling plants.

Plants divided on the basis of the process or stages in manufacturing: Each production process or stage of manufacturing may require distinctively different equipment capabilities, labor skills, technologies, and managerial policies and emphasis. Since the products of one plant feed into the other plant, this strategy requires much centralized coordination of the manufacturing activities from the corporate office that are expected to understand the various technological aspects of all the plants.

Plants emphasizing flexibility: This requires much coordination between plants to meet the changing needs and at the same time ensure efficient use of the facilities and resources. Frequent changes in the long-term strategy in order to improve be efficiently temporarily, are not healthy for the organization. In any facility location problem the central question is: Is this a location at which the company can remain competitive for a long time?

For an established organization in order to add on to the capacity, following are the ways:

Expansion of the facilities at the existing site: This is acceptable when it does not violate the basic business and managerial outlines, i.e., philosophies, purposes, strategies and capabilities. For example, expansion should not compromise quality, delivery, or customer service.

Relocation of the facilities (closing down the existing ones): This is a drastic step which can be called as "Uprooting and Transplanting". Unless there are very compelling reasons, relocation is not done. The reasons will be either bringing radical changes in technology, resource availability or other destabilization.

All these factors are applicable to service organizations, whose objectives, priorities and strategies may differ from those of hardcore manufacturing organizations.

III. In case of global location:

Because of globalization, multinational corporations are setting up their organizations in India and Indian companies are extending their operations in other countries. In case of global locations there is scope for virtual proximity and virtual factory.

Virtual Proximity: With the advance in telecommunications technology, a firm can be in virtual proximity to its customers. For a software services firm much of its logistics is through the information/ communication pathway. Many firms use the communications highway for conducting a large portion of their business transactions. Logistics is certainly an important factor in deciding on a location whether in the home country or abroad. Markets have to be reached. Customers have to be contacted. Hence, a market presence in the country of the customers is quite necessary.

Virtual Factory: Many firms based in USA and UK in the service sector and in the manufacturing sector often out sources part of their business processes to foreign locations such as India. Thus, instead of one's own operations, a firm could use its business associates operations facilities. The Indian BPO firm is a foreign-based company's virtual service factory'. So a location could be one's own or one's business associates. The location decision need not always necessarily pertain to own operations.

Reasons for a Global/Foreign Location:

A. Tangible Reasons:

The tangible reasons for setting up an operation facility abroad could be as follows:

Reaching the customer: One obvious reason for locating a facility abroad is that of capturing a share of the market expanding worldwide. The phenomenal growth of the GDP of India is a big reason for the multinationals to have their operations facilities in our country. An important reason is that of providing service to the customer promptly and economically which is logistics-dependent. Therefore, cost and case of logistics is a reason for setting up manufacturing facilities abroad. By logistics set of activities closes the gap between production of goods/services and reaching of these intended goods/services to the customer to his satisfaction. Reaching the customer is thus the main objective. The tangible and intangible gains and costs depend upon the company defining for it as to what that reaching means. The tangible costs could be the logistics related costs; the intangible costs may be the risk of operating is a foreign country. Tangible gains are the immediate gains; the intangible gains are an outcome of what the company defines the concepts of reaching and customer for it. The other tangible reasons could be as follows:

The host country may offer substantial tax advantages compared to the home country.

The costs of manufacturing and running operations may be substantially less in that foreign country. This may be due to lower labor costs, lower raw material cost, better availability of the inputs like materials, energy, water, ores, metals, key personnel etc.

The company may overcome the tariff barriers by setting up a manufacturing plant in a foreign country rather than exporting the items to that country.

B. Intangible Reasons:

The intangible reasons for considering setting up an operations facility abroad could be as follows:

1. Customer-related Reasons:

With an operations facility in the foreign country, the firm's customers may feel secure that the firm is more accessible. Accessibility is an important service quality determinant.

The firm may be able to give a personal tough.

The firm may interact more intimately with its customers and may thus understand their requirements better.

It may also discover other potential customers in the foreign location.

2. Organisational Learning-related Reasons:

The firm can learn advanced technology. For example, it is possible that cutting-edge technologies can be learning by having operations in a technologically more advanced country. The firm can learn from advanced research laboratories/universities in that country. Such learning may help the entire product-line of the company.

The firm can learn from its customers abroad. A physical location there may be essential towards this goal.

It can also learn from its competitors operating in that country. For this reason, it may have to be physically present where the action is.

The firm may also learn from its suppliers abroad. If the firm has a manufacturing plant there, it will have intensive interaction with the suppliers in that country from whom there may be much to learn in terms of modern and appropriate technology, modern management methods, and new trends in business worldwide.

3. Other Strategic Reasons:

The firm by being physically present in the host country may gain some _local boy' kind of psychological advantage. The firm is no more a foreign company just sending its products across international borders. This may help the firm in lobbying with the government of that country and with the business associations in that country.

The firm may avoid political risk by having operations in multiple countries.

By being in the foreign country, the firm can build alternative sources of supply. The firm could, thus, reduce its supply risks.

The firm could hunt for human capital in different countries by having operations in those countries. Thus, the firm can gather the best of people from across the globe.

Foreign locations in addition to the domestic locations would lower the market risks for the firm. If one market goes slow the other may be doing well, thus lowering the overall risk.

Factors influencing plant location:

Availability of raw materials:

It is essential for the organization to get raw material in right qualities and time in order to have an uninterrupted production. The cost obtaining Example Nearness to raw material is important in case of industries such as sugar, cement, jute and cotton textiles.

Nearness to market:

Every company is expected to serve its customers by providing goods and services at the time needed and at reasonable price organizations may choose to locate facilities close to the market or away from the market depending upon the product. When the buyers for the product are concentrated, it is advisable to locate the facilities close to the market. Locating nearer to the market is preferred if;

- The products are delicate and susceptible to spoilage.
- After sales services are promptly required very often.
- Transportation cost is high and increase the cost significantly.
- Shelf life of the product is low.

Nearness to the market ensures a consistent supply of goods to customers and reduces the cost of transportation.

Transportation facilities:

Speedy transport facilities ensure timely supply of raw materials to the company and finished goods to the customers. The transport facility is a prerequisite for the location of the plant. There are five basic modes of physical transportation, air, road, rail, water and pipeline. Goods that are mainly intended for exports demand a location near to the portor large airport. The choice of transport method and hence the location will depend on relative costs, convenience, and suitability. Thus transportation cost to value added is one of the criteria for plant location.

Climatic conditions:

The geology of the area needs to be considered together with climatic conditions (humidity, temperature). Climates greatly influence human efficiency and behaviour. Some industries require specific climatic conditions e.g., textile mill will require humidity.

Government policy:

The policies of the state governments and local bodies concerning labor laws, building codes, safety, etc., are the factors that demand attention. In order to have a balanced regional growth of industries, both central and state governments in our country offer the package of incentives to entrepreneurs in particular locations. The incentive package may be in the form of exemption from a sales tax and excise duties for a specific period, soft loan from financial institutions, subsidy in electricity charges and investment subsidy. Some of these incentives may tempt to locate the plant to avail these facilities offered.

Labour and wages:

The problem of securing adequate number of labor and with skills specific is a factor to be considered both at territorial as well as at community level during plant location. Importing labor is usually costly and involve administrative problem. The history of labor relations in a prospective community is to be studied. Prospective community is to be studied. Productivity of labor is also an important factor to be considered. Prevailing wage pattern, cost of living and industrial relation and bargaining power of the union's forms in important considerations.

Community infrastructure and amenity:

All manufacturing activities require access to a community infrastructure, most notably economic overhead capital, such as roads, railways, port facilities, power lines and service facilities and social overhead capital like schools, universities and hospitals.

Utilities, taxes, and real estate costs:

Other important factors that may emerge include utility costs (telephone, energy, and water), local and state taxes, financing incentives offered by local or state governments, relocation costs, and land costs.

Supporting industries and services:

Now a day the manufacturing organization will not make all the components and parts by itself and it subcontracts the work to vendors. So, the source of supply of component parts will be the one of the factors that influences the location.

Proximity to Customers:

Location is a key factor in determining how conveniently customers can carry on business with a firm. For example, few people would like to go to remotely located dry cleaner or supermarket if another is more convenient. Thus the influence of location on revenues tends to be the dominant factor.

Transportation Costs and Proximity to Markets:

For warehousing and distribution operations, transportation costs and proximity to markets are extremely important. With a warehouse nearby, many firms can hold inventory closer to the customer, thus reducing delivery time and promoting sales.

Capital:

By looking at capital as a location condition, it is important to distinguish the physiology of fixed capital in buildings and equipment from financial capital. Fixed capital costs as building and construction costs vary from region to region. But on the other hand buildings can also be rented and existing plants can be expanded. Financial capital is highly mobile and does not very much influence decisions.

Disposal of waste:

Some industries such as chemical plants leather industries, steel and plants etc, have the problem of disposal of effluents and the site selected should have provision for this availability of power is essential for any manufacturing firm coal and oil natural gas are sources of electric power in means a firm can gain one type of location benefit by giving up another.

Quality of life:

Good schools, recreational facilities, cultural events, and an attractive lifestyle contribute to quality of life. This factor is relatively unimportant on its own, but it can make the difference in location decisions.

Proximity to suppliers and resources:

In many companies, plants supply parts to other facilities or rely on other facilities for management and staff support. These require frequent coordination and communication, which can become more difficult as distance increases. There are some other factors needed to be considered, including room for expansion, construction costs, accessibility to multiple modes of transportation, the cost of shuffling people and materials between plants, competition from other firms for the workforce, community attitudes, and many others. For global operations, firms are emphasizing local employee skills and education and the local infrastructure

Various models are available which help to identify the ideal location. Some of the popular models are:

Factor rating method

Weighted factor rating method

Load-distance method

Centre of gravity method

Break even analysis

The process of selecting a new facility location involves a series of following steps:

Identify the important location factors.

Rate each factor according to its relative importance, *i.e.*, higher the ratings is indicative of prominent factor.

Assign each location according to the merits of the location for each factor.

Calculate the rating for each location by multiplying factor assigned to each location with basic factors considered.

Find the sum of product calculated for each factor and select best location having highest total score.

LOCATION PROBLEM

The plant location should be based on the company's expansion plan and policy, diversification plan for the products, changing market conditions, the changing sources of raw materials and many other factors that influence the choice of the location decision. The purpose of the location study is to find an optimum location one that will result in the greatest advantage to the organization.

ILLUSTRATION 1:

Let us assume that a new medical facility, Health-care, is to be located in Delhi. The location factors, factor rating and scores for two potential sites are shown in the following table. Which is the best location based on factor rating method?

Sl. No.	*Location factor*	*Factor rating*	*Rating*	
			Location 1	*Location 2*
1.	Facility utilization	8	3	5
2.	Total patient per month	5	4	3
3.	Average time per emergency trip	6	4	5
4.	Land and construction costs	3	1	2
5.	Employee preferences	5	5	3

SOLUTION:

Sl. No.	*Location factor*	*Factor rating (1)*	*Location 1*		*Location 2*	
			(Rating) (2)	*Total= (1) . (2)*	*(Rating) (3)*	*Total = (1) . (3)*
1.	Facility utilization	8	3	24	5	40
2.	Total patient per month	5	4	20	3	15
3.	Average time per emergency trip	6	4	24	5	30
4.	Land and construction costs	3	1	3	2	6
5.	Employee preferences	5	5	25	3	15
			Total	96	Total	106

Result:

The total score for location 2 is higher than that of location 1.

Hence location 2 is the best choice.

SOLUTION:

The weighted score for this particular site is calculated by multiplying each factors weight by its score and adding the results:

Weighed score location 1 = 25 × 3 + 25 × 4 + 25 × 3 + 15 × 1 + 10 × 5 = 75 + 100 + 75 + 15 +50=315

Weighed score location 2 = 25 × 5 + 25 × 3 + 25 × 3 + 15 × 2 + 10 × 3= 125 + 75 + 75 + 30 +30=335

Location 2 is the best site based on total weighted scores.

ILLUSTRATION 3:

From the following data select the most advantageous location for setting a plant for making transistor radios.

	Site X *Rs.*	*Site Y* *Rs.*	*Site Z* *Rs.*
(*i*) Total initial investment	2,00,000	2,00,000	2,00,000
(*ii*) Total expected sales	2,50,000	3,00,000	2,50,000
(*iii*) Distribution expenses	40,000	40,000	75,000
(*iv*) Raw material expenses	70,000	80,000	90,000
(*v*) Power and water supply expenses	40,000	30,000	20,000
(*vi*) Wages and salaries	20,000	25,000	20,000
(*vii*) Other expenses	25,000	40,000	30,000
(*viii*) Community attitude	Indifferent	Want business	Indifferent
(*ix*) Employee housing facilities	Poor	Excellent	Good

SOLUTION:

Total expenses	*Site X* *Rs.*	*Site Y* *Rs.*	*Site Z* *Rs.*
[Add (*iii*) (*iv*) (*v*) (*vi*) and (*vii*)]	1,95,000	2,15,000	2,35,000

$$\text{Rate of return (RoR), \%} = \frac{\text{Total sales} - \text{Total expenses}}{\text{Total investment}} \times 100$$

$$\text{RoR for Site X} = \frac{2,50,000 - 1,95,000}{2,00,000} \times 100$$

$$= \mathbf{27.5\%}$$

$$\text{RoR for Site Y} = \frac{3,00,000 - 2,15,000}{2,00,000} \times 100$$

$$= \mathbf{42.5\%}$$

$$\text{RoR for Site Z} = \frac{2,50,000 - 2,35,000}{2,00,000} \times 100$$

$$= \mathbf{7.5\%}$$

Location Y can be selected because of higher rate of return.

ILLUSTRATION 4:

The potential locations Chennai, Coimbatore and Madurai have the cost structure shown for producing telecommunication sets, expected to sell for Rs.90 find the best location for an expected volume of 1,850units /year.

SITE	FIXED COST/YEAR	VARIABLE COST/UNIT
Chennai	20,000	50
Coimbatore	40,000	30
Madurai	80,000	10

Solution:

Total cost = fixed cost +variable cost

Total cost at Chennai = 20,000 + (50*1,850) = 1, 12,500

Coimbatore = 40,000 + (30*1,850) = 95,000

Madurai = 80,000 + (10*1,850) = 98,500

Coimbatore is the best location profit = total revenue-total cost 90*1,850-95,500=71,000

RECENT TRENDS IN THE LOCATION OF INDUSTRIES

Priority for the suburban areas:

The industrialists show their preference for the suburban area as the site for establishment of a new unit or relocation of the existing one. The industrial policy of the government does not permit the establishment of a new unit or expansion of an existing one in city areas. At the same time infrastructure facilities are developed in the sub urban areas.

Industrial development in the notified backward areas:

In order to have balanced regional development, the Central Government as well as the State Government has notified certain backward areas Different types of incentives like cash subsidy, tax relief, financial assistance with low interest rates, cheaper land and power supply etc are provided. So, many such areas have been developed substantially in the recent times.

Decentralization of industries:

Under the conscious industrial policy of the Government, concentration of industrial units is prevented through licensing policy. New units are not permitted to be started in certain industrially congested areas. Similarly, existing units either establish their additional plants in a less developed area or sometimes relocate the whole unit in such areas.

Competition between government and institutions:

As industry provides job opportunities to the local population, many local organizations attempt to tempt the prospective promoters to establish the units in their areas. They provide different types of incentives like cheap land, relief in local taxes etc. Sometimes the objective of local organizations and the government comes in conflict on the issues of location of industries. Thus, the whole pattern of decision about the location of industries has undergone substantial changes in recent times.

PLANT LAYOUT - CONCEPT

After deciding above the proper site for locating an industrial unit, next important point to be considered by an entrepreneur is to decide about the appropriate layout for the plant. Plant layout is primarily concerned with the internal set up of an enterprise in a proper manner. The concept of plant layout is not static but dynamic one. It is on account of continuous manufacturing and technological improvements taking place necessitating quick and immediate changes in production processes and designs. A new layout may be necessary because of technological changes in the products as well as simple change in processes, machines, methods and materials".

"Plant layout is the arrangement of machines, work areas and service areas within a factory".

- George R. Terry.

"Plant layout involves the development of physical relationship among building, equipment and production operations, which will enable the manufacturing process to be carried on efficiently". - Morris E. Hurley.

Objectives of plant layout:

The primary objective of plant layout is to maximize production at minimum cost. The layout should be designed in such a way that it is flexible to change according to new processes and production techniques. The layout should be able to satisfy the needs of all those who are associated with the production system such as workers, supervisors, managers etc., to fulfill the above goals, the plant layout should be designed with the following objectives:

Minimizing handling of materials.

Maintaining flexibility of operations.

Ensuring optimum utilization of men, materials, equipment and available space.

Achieving good work flow and avoiding accumulation of work.

Minimizing delays and bottlenecks in the production system.

Ensuring safety of workmen by minimizing and eliminating the chances of accidents.

Providing for effective supervision and production control.

Minimizing work-in-process inventory.

Providing sufficient and conveniently located service centres.

Flexibility in design to adapt to the changing future requirements.

Need of Plant layout:

Many situations give rise to the problem of plant layout. Two plants having similar operations may not have identical layout. This may be due to size of the plant, nature of the process and management's caliber. The necessity of plant layout may be feeling and the problem may arise when.

There are design changes in the product.

There is an expansion of the enterprise.

There is proposed variation in the size of the departments.

Some new product is to be added to the existing line.

Some new department is to be added to enterprise and there is reallocation of the existing department.

A new plant is to be set up.

Principles of Plant Layout:

While designing the plant layout, the following principles must be kept in view:

Principle of minimum movement: Materials and labor should be moved over minimum distances; saving cost and time of transportation and material handling.

Principle of space utilization: All available cubic space should be effectively utilized – both horizontally and vertically.

Principle of flexibility: Layout should be flexible enough to be adaptable to changes required by expansion or technological development.

Principle of interdependence: Interdependent operations and processes should be located in close proximity to each other; to minimize product travel.

Principle of overall integration: All the plant facilities and services should be fully integrated into a single operating unit; to minimize cost of production.

Principle of safety: There should be in-built provision in the design of layout, to provide for comfort and safety of workers.

Principle of smooth flow: The layout should be so designed as to reduce work bottlenecks and facilitate uninterrupted flow of work throughout the plant.

Principle of economy: The layout should aim at effecting economy in terms of investment in fixed assets.

Principle of supervision: A good layout should facilitate effective supervision over workers.

Principle of satisfaction: A good layout should boost up employee morale, by providing them with maximum work satisfaction.

Types of Plant Layout:

Layouts can be classified into the following five categories:

Product layout

Process layout

Combination layout

Fixed position layout

Service Facility layout

Product Layout (or Line Layout):

In this type of layout, all the machines are arranged in the sequence, as required to produce a specific product. It is called line layout because machines are arrange in a straight line. The raw materials are fed at one end and taken out as finished product to the other end. The raw material is supplied at one end of the line and goes from one operation to the next quite rapidly with a minimum work in process, storage and material handling.

Advantages offered by Product Layout:

(i) Lowers total material handling cost.

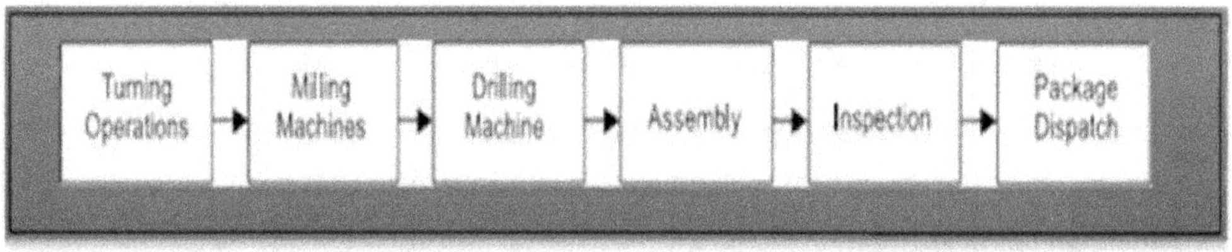

(ii) There is less work in processes.

(iii) Better utilization of men and machines,

(iv) Less floor area is occupied by material in transit and for temporary storages.

(v) Greater simplicity of production control.

(vi) Total production time is also minimized.

Limitations of Product Layout:

(i) No flexibility which is generally required is obtained in this layout.

(ii) The manufacturing cost increases with a fall in volume of production.

Process or Functional Layout:

The process layout is particularly useful where low volume of production is needed. If the products are not standardized, the process layout is lower desirable, because it has creator process flexibility than other. In this type of layout, the machines and not arranged according to the sequence of operations but are arranged according to the nature or type of the operations. This layout is commonly suitable for non repetitive jobs.

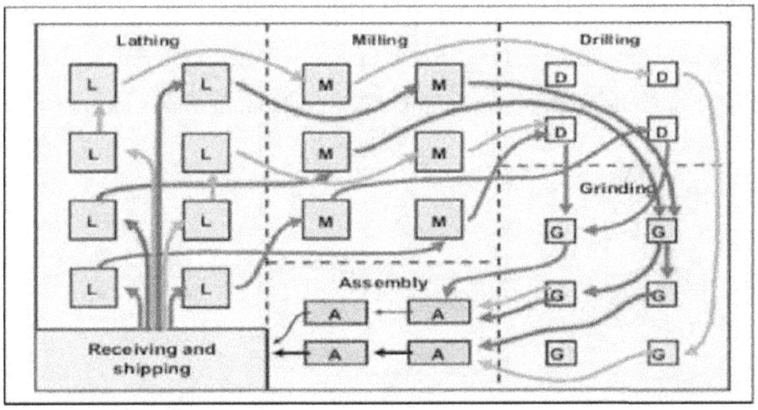

Same type of operation facilities are grouped together such as lathes will be placed at one place; all the drill machines are at another place and so on. Therefore, the process carried out in that area is according to the machine available in that area.

Advantages of Process Layout:

(i) There will be less duplication of machines. Thus, total investment in equipment purchase will be reduced.

(ii) It offers better and more efficient supervision through specialization at various levels.

(iii) There is a greater flexibility in equipment and man power thus load distribution is easily controlled.

(iv) Better utilization of equipment available is possible.

(v) Break down of equipment can be easily handled by transferring work to another machine/work station.

(vi) There will be better control of complicated or precision processes, especially where much inspection is required.

Limitations of Process Layout:

(i) There are long material flow lines and hence the expensive handling is required.

(ii) Total production cycle time is more owing to long distances and waiting at various points.

(iii) Since more work is in queue and waiting for further operation hence bottle necks occur.

(iv) Generally, more floor area is required.

(v) Since work does not flow through definite lines, counting and scheduling is more tedious.

(vi) Specialization creates monotony and there will be difficult for the laid workers to find job in other industries.

Fixed Position Layout:

This type of layout is the least important for today's manufacturing industries. In this type of layout the major component remain in a fixed location, other materials, parts, tools, machinery, man power and other supporting equipment's are brought to this location.

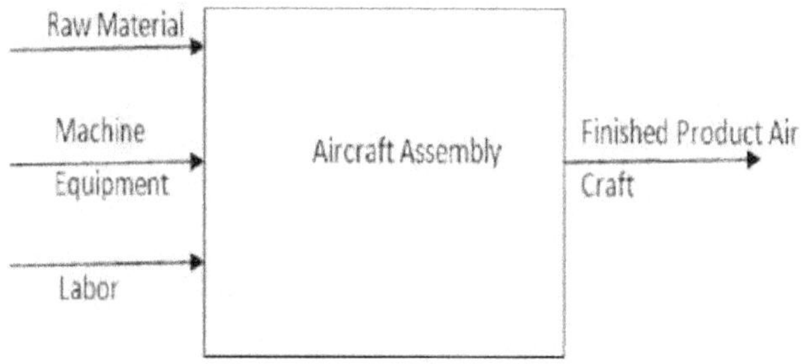

The major component or body of the product remain in a fixed position because it is too heavy or too big and as such it is economical and convenient to bring the necessary tools and equipment's to work place along with the man power. This type of layout is used in the manufacture of boilers, hydraulic and steam turbines and ships etc.

Advantages of Fixed Position Layout:

(i) Material movement is reduced.

(ii) Capital investment is minimized.

(iii) The task is usually done by gang of operators, hence continuity of operations is ensured.

(iv) Production centers are independent of each other. Hence, effective planning and loading can be made. Thus total production cost will be reduced.

(v) It offers greater flexibility and allows change in product design, product mix and production volume.

Limitations of Fixed Position Layout:

(i) Highly skilled man power is required.

(ii) Movement of machines equipment's to production centre may be time consuming.

(iii) Complicated fixtures may be required for positioning of jobs and tools. This may increase the cost of production.

Combination Type of Layout:

Now a day in pure state any one form of layouts discussed above is rarely found. Therefore, generally the layouts used in industries are the compromise of the above mentioned layouts. Every layout has got certain advantages and limitations. Therefore, industries would to like use any type of layout as such. Flexibility is a very important factory, so layout should be such which can be molded according to the requirements of industry, without much investment. If the good features of all types of layouts are connected, a compromise solution can be obtained which will be more economical and flexible.

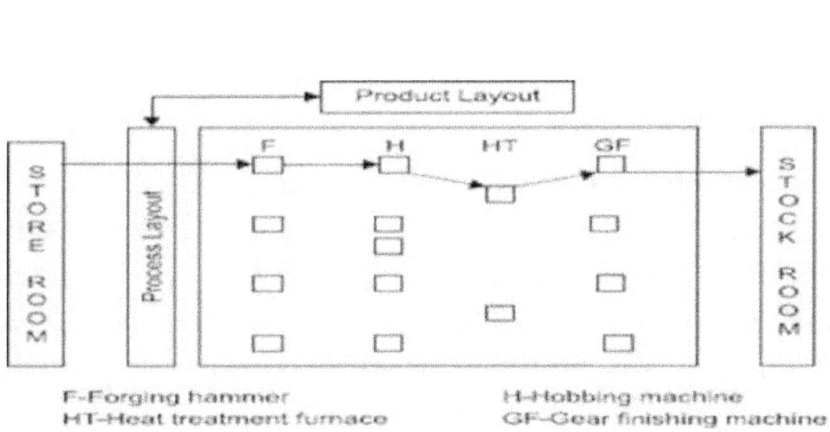

In practice, plants are rarely laid out either in product or process layout form. Generally a combination of the two basic layouts is employed; to derive the advantages of both systems of layout. For example, refrigerator manufacturing uses a combination layout.

Advantages:

Component standardization rationalization.

Effective machine operation and productivity.

Customer service.

It can decrease the paper work and overall production time. Work-in-progress and work movement

Service Facility Layout:

Services facility layouts should provide for easy entrance to these facilities Large, well organized and well-designed walk ways to and from parking areas are some of the requirements of service requirements The fundamental difference between service facility and manufacturing facility layouts is that many service facilities exist together customers and services. Examples; Hospitals – processing of physical materials and production efficiency Banks - customer receiving and servicing function.

LAYOUT DESIGN PROCEDURE:

Layout design procedures can be classified into manual methods and computerized methods.

Manual methods: Under this category, there are some conventional methods like travel chart and Systematic Layout Planning (SLP).

Computerized methods: Under this method, again the layout design procedures can be classified in to constructive type algorithm and improvement type algorithms. Construction type algorithms Automated Layout Design program (ALDEP) Computerized Relationship

Layout Planning (CORELAP) Improvement type Algorithm Computerized Relative Allocation of Facilities Technique (CRAFT).

Computerized Relative Allocation of Facilities Technique (CRAFT):

CRAFT algorithm was originally developed by Armour and Buffa. It is an improvement algorithm. It starts with an initial layout and improves the layout by interchanging the departments pairwise so that the transport cost is minimized. The algorithm continues until not further interchanges are possible to reduce the transportation cost. The result given by CRAFT is not optimum in terms of minimum cost of transportation. But the result will be good and close to optimum in majority of applications. Hence, CRAFT is mainly a heuristic algorithm. Unfortunately, plant layout problem comes under combinatorial category. So, usage of efficient heuristic like CRAFT is inevitable for such problem.

Features of CRAFT:

The major features of CRAFT are as listed below:

Attempts to minimize transportation cost, where **transportation cost = flow x distance x unit cost**. Required the assumptions that:

Move costs are independent of the equipment utilization and

Move costs are linearly related to the length of the move.

Distance matrix used in the rectilinear distance between department centroids. CRAFT being a path-oriented method, the final layout is dependent on the initial layout. Therefore, a number of different initial layouts should be used as input to the CRAFT layout. CRAFT allows the use of dummy departments to represent fixed areas in the layout.

CRAFT input requirements are:

Initial layout.

Flow data.

Cost per unit distance.

Total number of departments.

Fixed departments and their location. Area of departments.

CRAFT Procedure:

The steps of CRAFT algorithm are summarized below:

Step1. Input: 1. Number of departments

2. Number of interchangeable departments
3. Initial layout
4. Cost matrix
5. Flow matrix (Load summary)
6. Area of departments.

Step2. Compute centroids of departments in the present layout.

Step3. Form distance matrix using the centroids.

Step4. Given data on flow, distance and cost, compute the total handling cost of the present layout.

Step5. Find all the possible pairwise interchanges of departments based on common border or equal area criterion. For each possibility, interchange the corresponding centroids and compute approximate costs.

Step6. Find the pair of departments corresponding to the minimum handling cost from among all the possible pairs of interchanges.

Step7. Is the cost in the previous step less than the total cost of the present layout? If yes, go to step 8. If not, go to step 11.

Step8. Interchange the selected pair of departments. Call this as the NEW LAYOUT. Compute centroids, distance matrix and total cost.

Step9. Is the cost of new layout less than the cost of the present layout? If yes, go to Step 10. If not, go to step 11.

Step10. The new layout is here after considered as the PRESENT LAYOUT. Its data on centroids, layout matrix and the total cost is retained. Go to step5. Step11. Print the present layout as the FINAL LAYOUT.

Step12. Stop.

AUTOMATED LAYOUT DESIGN PROGRAM (ALDEP)

Now we will examine Automated Layout Design Program (ALDEP).ALDEP is basically a construction algorithm but it can also be used to evaluate two layouts. The algorithm uses basic data on facilities and builds a layout by successively placing the layout using relationship information between the departments. The basic inputs to ALDEP are;

Length and width of facility. Area of
each department.
Minimum closeness preference (MCP) value. Sweep
width.
Relationship chart showing the closeness rating.
Location and size of any restricted area.

The procedures adopted for using ALDEP are:

Step 1: Input the following: 1. Length and width of facility.
2. Area of each department.
3. Minimum closeness preference (MCP) value.
4. Sweep width.
5. Relationship chart showing the Closeness rating.
6. Location and size of restricted area.

Step 2: One department is selected randomly and placed in the layout.

Step 3: In this step, the algorithm uses minimum closeness required between departments for the selection of departments to be placed with an earlier placed department. Select the department having maximum closeness rating. If there is no department having minimum closeness preference, then any department that remains to be placed is selected. Step 4: If all the departments are placed in the layout, go to step 5. Else, go to step 3.

Step 5: Compute the total score of the layout.

Step 6: If the total score required is the acceptable score, then go to step7, else go tostep2.

Step 7: Print the current layout and the corresponding score.

Computerized Relationship Layout Planning (CORELAP)

This algorithm is based on Muther's procedure given in systematic Layout Planning. A computer algorithm was developed by R.C. Lee. Interactive version was developed by James Moore.

Input requirements;

Number of departments and their area.
Closeness relationship as given by REL-chart.
Weighted rating for REL-chart entries. Optional input information Scale
of output.

Building length to width ratio.

Department pre-assignment.

General approach is to select the most critical department first, and place it at the centre of the layout. After the first department is placed, then the department having highest closeness relationship with the department which is already placed is selected and placed in the best location adjacent to the previously placed departments. CORELAP builds the layout from centre. The final layout will not have a regular rectangular shape. The user has to modify it slightly to suit the situation. Final score of the layout is developed by using the closeness values and rectilinear distances between all pairs of the departments.

CORELAP algorithm: **the following are the major steps of CORELAP algorithm.**

Defining basic data.

Determination of placement of order.

Placement of departments in the layout. Finding

the total score of the layout.

LINE BALANCING is useful tool. Line-balancing strategy is to make production lines stretchy enough to absorb external and internal indiscretion. This strategy involves setting a planned rate of production for necessary materials to be fabricated within a particular time frame.

Additionally, successful line balancing requires assuring that every line segment's production quota can be met within the time frame using the available production capacity. This is an efficient device to develop the throughput of assembly lines and work cells while decreasing manpower requirements and expenses. Line-balancing is slightly different from assembly line balancing. There are two types of line balancing that include Static Balance and Dynamic Balance. Static Balance denotes long-term differences in capacity over a period of several hours or longer. Static imbalance results in underutilization of workstations, machines and people. Dynamic Balance refers to short-term differences in capacity such as over a period of

minutes, hours at most. Dynamic imbalance occurs from product mix changes and difference in work time dissimilar to product mix. The intent of Line balancing is to match the output rate to the production plan. This will help organization to make sure on-time delivery and avoids build-up of surplus inventory.

Line balancing operates under two circumstances:

Precedence Constraint: Products cannot progress to other station if it doesn't complete necessary task at that station. It should not across other station because certain part needs to be performed before other activities.

Cycle time Restriction: Cycle time is maximum time for products spend in every workstation. Different workstation has different cycle time.

Objective of Line Balancing: Following are major objectives of Line balancing procedure. It is used to:

Manage the workloads among assemblers.

Recognize the location of bottleneck.

Decide number of workstation. Decrease

production cost.

Assigning task to each work station in such a way that there is little idle time.

Steps in Solving Line Balancing

There are four steps in solving line balancing

a. **Drawing Precedence Diagram**: Precedence diagram needs to be drawn to demonstrate a relationship between workstations. Certain process begins when previous process was done.

b. **Determining Cycle Time:** Cycle time is longest time allowed at each station. This can be expressed by this formula:

$$\textbf{Cycle time} = \frac{\textbf{Available time}}{\textbf{Desired output}}$$

This means the products need to leave the workstations before it reaches its cycle time.

c. **Assigning tasks to workstation**: The tasks distributions should be taken after completing a time cycle. It's good to allocate tasks to workstation in the order of longest task times.

$$\textbf{Number of work Stations} = \frac{\sum \textbf{ Task Time}}{\textbf{Desired Actual Time}}$$

d. Calculating an Efficiency Line: This is done to find effectiveness of the line. The formula is given by:

$$\textbf{Line Efficiency} = \frac{\textbf{Sum of task times}}{\textbf{Number of workstation X Desired cycle time}}$$

ASSEMBLY LINE BALANCING involved the action of assembles different parts together. It involves many production lines while normal Line-balancing may only involve one production line. Assembly Line Balancing is the problem of assigning operations to workstations along an assembly line, in such a way that the assignment be best in some sense. Since introduction of assembly lines by Henry Ford, Line-balancing has been an optimization problem of important industrial importance. The efficiency difference between an optimal and a sub-optimal assignment can yield economies reaching huge amount per year. Line balancing technique was used normally in assembly line of the automotive industry which is called ALB. Most of the Small and Medium Industries do not use line balancing method in the production line.

What is Ranked Positional Weight Method: It is a method used for line development and balancing. It takes into account the precedence relationships as well as processing time of all tasks.

Ranked Positional Weight (RPW)

Step 1 - Calculate the weight of the position of each work elements.

Step 2 - Sort the elements according to the weight of the position of the largest to the smallest.

Step 3 – calculate the cycle time.

Step 4 – place the working element with the greatest weight on all the workstations do not violate precedence relationships and time station does not exceed the cycle time. Step 5 – Repeat step 4, until all the elements are placed.

Step 6 - After forming a work station consisting of its elements, and then specify the value of efficient, balance delay, and its smoothest index.

COMSOAL: Today's highly competitive market influences the manufacturing industry to improve their production systems to become the optimal system in the shortest cycle time as possible. One of most common problems in manufacturing systems is the assembly line balancing problem. The assembly line balancing problem involves task assignments to workstations with optimum line efficiency. The line balancing technique, namely "COMSOAL", is an abbreviation of "Computer Method for Sequencing Operations for Assembly Lines". Arcus initially developed the COMSOAL technique in 1966, and it has been mainly applied to solve assembly line balancing problems. The most common purposes of COMSOAL are to minimize idle time, optimize production line efficiency, and minimize the number of workstations.

Objectives that should be gained balancing an assembly line are as follows;

Regular material flow.

Maximum usage of man power and machine capacity.

Minimum process times.

Minimizing slack times

Minimizing workstations.

Distribute slack times to workstations. Reduce production costs.

COMSOAL - PROCEDURE

1. The first step is creating the table, which presents all activities lists in order, considering by precedence relationships.
2. Secondly, selecting the available activities from the table that has no predecessor task, in other words all predecessor tasks of considering activity need to be finished.
3. Creating the available activities list.
4. Choosing activities from the available lists to the workstation until the total processing time of all activities in the workstation is nearly or equal to the given cycle time.
5. The next step is recreating the new available activity list.
6. Repeating steps 2-5 until all activities are assigned into workstations.
7. The final step is keeping the possible solution and then repeating steps 1-5 to find the alternative solution, until the best solution is obtained.

OPERATIONS MANAGEMENT – QUESTION BANK

UNIT – III

PART – A

S. No	Questions (5 marks)	CO	Level
1	Explain the objectives of plant location.	CO3	L2
2	Explain the recent trends in the location of industry.	CO3	L4
3	List out the plant layout benefits.	CO3	L4
4	Bring out features of CORELAP.	CO3	L3

<table>
<tr><td>5</td><td>TVS group of companies; manufactures of auto spares choose three factors and four sites.
<table><tr><td rowspan="2">Factor</td><td colspan="4">Potential sites</td></tr><tr><td>S1</td><td>S2</td><td>S3</td><td>S4</td></tr><tr><td>F1</td><td>3</td><td>6</td><td>8</td><td>2</td></tr><tr><td>F2</td><td>4</td><td>5</td><td>9</td><td>3</td></tr><tr><td>F3</td><td>7</td><td>2</td><td>6</td><td>3</td></tr></table>
The sites were assigned ratings between 0-10 points against each factor; the sum of the site ratings was used to compare it with other sites.</td><td>CO3</td><td>L3</td></tr>
<tr><td>6</td><td>List out the advantages for process layout.</td><td>CO3</td><td>L2</td></tr>
<tr><td>7</td><td>Recall a short note on ranked positional weight method.</td><td>CO3</td><td>L3</td></tr>
<tr><td>8</td><td>Brief fixed position layout.</td><td>CO3</td><td>L3</td></tr>
<tr><td>9</td><td>“Layout expansion is very easy compared to layout revision”. Explain.</td><td>CO3</td><td>L4</td></tr>
<tr><td>10</td><td>Comsoal- give abbreviation and its objectives.</td><td>CO3</td><td>L4</td></tr>
</table>

PART – B

S. NO	Questions (10 marks)	CO	Level
1	Illustrate and explain the types of layout.	CO3	L5
2	Summarizes factors affecting in quantitative factors in plant location.	CO3	L4
3	Analysis the merits and demerits of locating a plant in sub urban area.	CO3	L4
4	Elucidate the procedure for CRAFT.	CO3	L4
5	Enumerate the procedure adopted for using ALDEP.	CO3	L5
6	Elaborate line balancing operates in two different circumstances.	CO3	L5

PART – C

CASE STUDY – (20 marks)

S. No	Questions (20 marks)	CO	Level
1	Alpha a four-wheeler company is a leading company in the south manufacturing chassis of bus/lorry in 600 acres of land with 3000 employees. The annual production capacity of the plant is 6000 chassis; The market research department projected its future demand to be 2.5 times the present capacity of the plant. So, the company took a decision to set up another plant in the north with a capacity of 75,000 chassis. It is in the process of procurement of the required land of 1,000 acres. The projected number of employees in the new factory would be 4,000. The productivity of any company mainly depends on the type of layout that is used to carry-out the activities to produce the product. so, the industrial engineering department of the exiting company is given the task of design the righty type of layout for the new company, All the sections of the automobile company will not have the same type of layout. The final assembly of chassis is done on a powered conveyer belt. This part of the company uses product layout which assembles the necessary subassemblies and components to form a full chassis. Questions: a .What do you suggest to Alpha company about this problem ?	CO3	L6

<table>
<tr><td></td><td>b. What is suitable layout for Alpha company to extend?</td><td></td><td></td></tr>
<tr><td>2</td><td>A company has to decide on location of a new plant. It has three locations A, B and C whose data is furnished below. Use a suitable
<table>
<tr><th></th><th colspan="3">Locations</th></tr>
<tr><th>Data</th><th>A (Rs.)</th><th>B (Rs.)</th><th>C(Rs.)</th></tr>
<tr><td>Wages and salaries</td><td>20,000</td><td>20,000</td><td>20,000</td></tr>
<tr><td>Power & water supply</td><td>20,000</td><td>30,000</td><td>25,000</td></tr>
<tr><td>Raw materials</td><td>80,000</td><td>75,000</td><td>60,000</td></tr>
<tr><td>Total initial investment</td><td>2,00,000</td><td>3,00,000</td><td>2,50,000</td></tr>
<tr><td>Distribution expenses</td><td>50,000</td><td>40,000</td><td>60,000</td></tr>
<tr><td>Miscellaneous expenses</td><td>40,000</td><td>25,000</td><td>30,000</td></tr>
<tr><td>Expected sales per year</td><td>2,25,000</td><td>2,50,000</td><td>2,25,000</td></tr>
</table>
criterion to advice the company on the best choice.</td><td>CO3</td><td>L6</td></tr>
</table>

TEXT / REFERENCE BOOKS

R. Paneerselvam, Production and Operations Management, PHI Learning Private Ltd, 3rd edition, 2012.

R. B. Kanna, Production and Operations Management, PHI Learning Private Ltd, 2nd edition, 2015.

P. Saravanavel & S.Sumathi, Production and Materials Management, Margham Publications, 2nd edition, 2012.

S. N. Chary, Production and Operations Management, Tata McGraw Hill Education Private Limited, 4th edition, 2009.

S. P. Singh, Production and Operations Management, Vikas Publishing House Private Limited, 1st edition, 2014.

IV. QUALITY CONTROL

Quality Control – Objectives, Importance, Quality Control Techniques – Control Charts - $\bar{x}$ Chart, R Chart, P Chart, C Chart – Acceptance Sampling – Work Study – Method Study, Time Study - Human factors in job design – Ergonomics – Work Environment and Workers Safety.

QUALITY CONTROL - CONCEPT

Quality does not mean the quality of manufactured product only. It may refer to the quality of the process (i.e., men, material, and machines) and even that of management. Where the quality manufactured product referred as or defined as, "Quality of product as the degree in which it fulfills the requirement of the customer. It is not absolute but it judged or realized by comparing it with some standards".

Quality begins with the design of a product in accordance with the customer specification further it involved the established measurement standards, the use of proper material, selection of suitable manufacturing process etc., quality is a relative term and it is generally used with reference to the end use of the product.

Crosby defined as, "Quality is conformance to requirement or specifications".

Juran defined as, "Quality is fitness for use". The Quality of a product or service is the fitness of that product or service for meeting or exceeding its intended use as required by the customer.

Quality Control (QC) may be defined as, "a system that is used to maintain a desired level of quality in a product or service". It is a systematic control of various factors that affect the quality of the product. It depends on materials, tools, machines, type of labor, working conditions etc.

QC is a broad term, it involves inspection at particular stage but mere inspection does not mean QC. As opposed to inspection, in quality control activity emphasis is placed on the quality future production. Quality control aims at prevention of defects at the source, relies on effective feedback system and corrective action procedure. Quality control uses inspection as a valuable tool.

According to Juran, "Quality control is the regulatory process through which we measure actual quality performance, compare it with standards, and act on the difference".

Another definition of quality control is from ANSI/ASQC standard (1978) quality control is defined as, "The operational techniques and the activities which sustain a quality of product or service that will satisfy given needs; also the use of such techniques and activities".

Alford and Beatty define QC as, "In the broad sense, quality control is the mechanism by which products are made to measure up to specifications determined from customers, demands and transformed into sales engineering and manufacturing requirements, it is concerned with making things right rather than discovering and rejecting those made wrong".

Objectives of Quality Control:

The objectives of quality control may be stated as follows:

1. To take the necessary corrective steps to keep the quality of the product from dropping below the desired level during manufacturing.

2. To analyze the trend and extent of quality deviation in a part or product during manufacture and to determine the cause of such deviation by statistical techniques when it cannot reasonably be attributed to chance.

3. To avoid as far as reasonably possible, having products reach the customer which are of lower quality than that which is considered acceptable.

4. To establish standards of quality those are readily acceptable to the customer and economical to maintain.

Some of the importance or benefits of quality control are:

Encourages quality consciousness

Satisfaction of consumers

Reduction in production cost

Most effective utilization of resources.

Reduction in inspection costs

Increased goodwill

Higher morale of employees

Improved employer-employee relations

Improved techniques and methods of production

Facilitates price fixation

Increased sales

STATISTICAL QUALITY CONTROL - CONCEPT

In the highly competitive market today, the main objective of manufacturers or producers is to achieve quality assurance in manufacturing and service organizations. In order to achieve this objective, different statistical tools have been developed, which are useful for controlling the quality of products vis-avis the specifications or standard. The technique of using statistical tools for controlling product quality vis-a-vis the specifications is known as Statistical Quality Control (SQC).

Statistical quality control is defined as the technique of applying statistical methods based on the theory of probability and sampling to establish quality standard and to maintain it in the most economical manner.

Elements of SQC:

The following are the main elements of SQC:

a) Sample Inspection: We know that 100% inspection needs huge expenditure of time, money, labor and resources. Further, if the nature of the product is such that it is completely destroyed during the process of inspection, e.g., a bulb, candle, ammunition, food, etc., 100% inspection is not practicable. Therefore, SQC is based on sampling inspection. In sampling inspection method, some items or units (called sample) are randomly selected from the process and then each and every unit of the sample is inspected.

b) Use of Statistical Methods: Some commonly used statistical tools such as random sampling, mean, range, standard deviation, mean deviation, standard error and concepts such as probability, binomial distribution, Poisson distribution, normal distribution, etc., are used in SQC. Since, quality control method involves extensive use of statistics; it is termed as Statistical Quality Control.

c) Fundamental Objective: The fundamental objective of SQC is to decide whether the unit produced is according to its specifications or not. If the unit produced is not according to its specifications and there is a variation in quality, it becomes necessary to trace the causes of variation and eliminate them if possible.

d) Decision Making: With the help of SQC, we decide whether the quality of the product or the process of manufacturing/producing goods is under control or not.

e) Specifications, Production and Inspection: SQC method helps in deciding about the specifications, production and inspection of a product.

Techniques of Statistical Quality Control:

The important techniques used for statistical quality control can be broadly classified into two categories:

Statistical Process Control (SPC) or simply Process Control,

and Product Control.

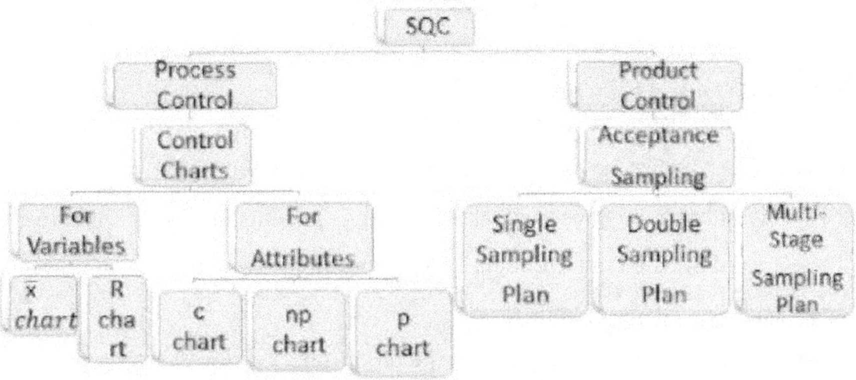

CONTROL CHARTS

It distinguishes special causes of variations from common causes of variation. They are used to monitor and control process on an ongoing basis. A typical control chart plots a selected quality characteristic found from sub-group of observations as a function of sample number. Characteristics such as sample average, sample range and sample proportion of non-conforming units are plotted. The centre line on a control chart represents the average value of characteristics being plotted. Two limits know as the upper control limit (UCL) and lower control limit (LCL) are also shown on control charts. These limits are constructed so that if the process is operating under a stable system of chance causes, the problem of an observation falling outside these limits is quite small.

Control chart shows the performance of a process from two points of view. First, they show a snapshot of the process at the moment the data are collected. Second, they show the process trend as time progresses. Process trends are important because they help in identifying the out of- control status if it actually exists. Also, they help to detect variations outside the normal operational limits, and to identify the cause of variations.

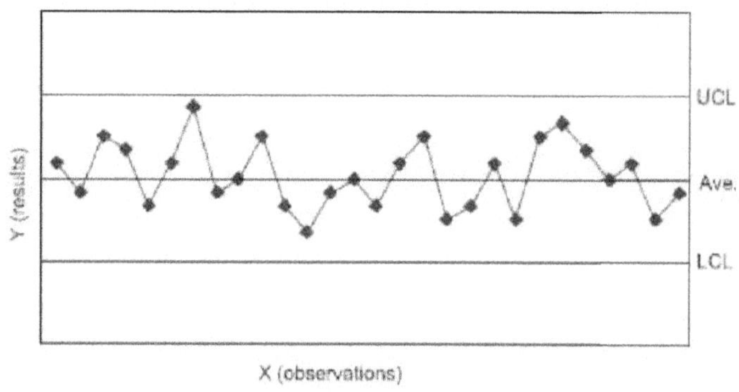

Objectives of Control Charts:

To secure information to be used in establishing or changing specifications or in determining whether the process can meet specifications or not.

To secure information to be used on establishing or changing production procedures.

To secure information to be used on establishing or changing inspection procedures or acceptance procedures or both.

To provide a basis for current decision during production.

To provide a basis for current decisions on acceptance for rejection of manufacturing or purchased product.

To familiarize personnel with the use of control chart.

Characteristics of Control Charts:

A control chart is a time-ordered diagram to monitor a quality characteristic, consisting of:

A nominal value, or centre line, the average of several past samples.

Two control limits used to judge whether action is required, an upper control limit (UCL) and a lower control limit (LCL).

Data points, each consisting of the average measurement calculated from a sample taken from the process, ordered overtime. By the Central Limit Theorem, regardless of the distribution of the underlying individual measurements, the distribution of the sample means will follow a normal distribution. The control limits are set based on the sampling distribution of the quality measurement

Benefits of Using Control Charts:

Following are the benefits of control charts:

A control chart indicates when something may be wrong, so that corrective action can be taken.

The patterns of the plot on a control chart diagnosis possible cause and hence indicate possible remedial actions.

It can estimate the process capability of process.

It provides useful information regarding actions to take for quality improvement.

ACCEPTANCE SAMPLING:

This is another technique of statistical quality control. This is also referred as 'Sampling Inspection plan.' This method is usually followed after goods have been produced or are in the final stage of production. Thus, it can be said that it is a post mortem of the quality of the product that has already been produced.

Under this method, a sample of the product produced is selected at random to study in detail whether the product conforms to the pre-determined standards or not. A limited percentage of defective products are allowed. But it has been observed that sometimes the sample selected turns out to be good, but the lot represented by the sample may be defective or sub-standard. In order to have more accurate and exact results, more than one sample of the product should be selected for carrying out the Sampling Inspection Plan.

The technique Acceptance Sampling undertakes two limiting levels of quality namely;

The Acceptable Quality Level (AQL) i.e. the least number or percentage of defective products that the buyer expects to purchase and the seller expects to sell and

The Lot Percentage Tolerance Defective (LPTD) refers to that limit where the buyer wants to be certain about the rejection of the lot.

This technique can be greatly helpful for improving relations between vendor and the customer which may be adverse on account of disputes relating to quality. Both the parties may sit together and mutually decide the limits within which quality should be accepted.

Advantages of Statistical Quality Control:

Following are the important benefits derived from the technique of statistical qualify control:

Lesser cost of inspection: Statistical quality control is based on sampling technique which involves lesser cost of inspection thereby cost of production is considerably reduced.

Increase in profits: By minimizing rejections, statistical quality control ensures the production of standard products which bring higher profits for the producer.

Setting tolerance limits: Quality control charts clearly lay down the tolerances limits beyond which the product is to be rejected. The results shown by these charts are more authentic and correct.

Develops quality consciousness: Statistical quality control is greatly helpful in developing the feeling of quality consciousness among the workers working in an organisation. This improves their functioning and reduces the number of defective operations undertaken by them,

Enhances reputation of the concern: By adopting the techniques of statistical quality control, pre-determined quality of the product is achieved and consumers get desired quality products. This brings good name to the firm and increases its goodwill among the people.

Improved relations between vendor and customers: It is greatly helpful in improving relations between supplier and the purchaser of material, by clearly fixing the tolerance limits with regard to quality of the goods supplied. This minimizes the possibility of any dispute between both the parties.

WORK STUDY - CONCEPT

Work study is a generic term for those techniques, method study and work measurement which are used in the examination of human work in all its contexts. And which lead systematically to the investigation of all the factors which affect the efficiency and economy of the situation being reviewed, in order to effect improvement."

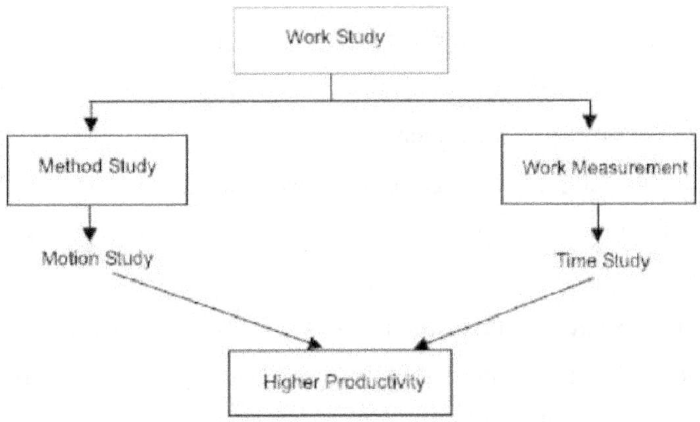

Work study is a means of enhancing the production efficiency (productivity) of the firm by elimination of waste and unnecessary operations. It is a technique to identify non-value adding operations by investigation of all the factors affecting the job. It is the only accurate and systematic procedure oriented technique to establish time standards. It is going to contribute to the profit as the savings will start immediately and continue throughout the life of the product. Method study and work measurement is part of work study. Part of method study is motion study; work measurement is also called by the name 'Time study'.

Advantages of work study:

It helps to achieve the smooth production flow with minimum interruptions.

It helps to reduce the cost of the product by eliminating waste and unnecessary operations.

Better worker-management relations.

Meets the delivery commitment.

Reduction in rejections and scrap and higher utilization of resources of the organization.

Helps to achieve better working

conditions. Better workplace layout.

Improves upon the existing process or methods and helps in standardization and Simplification.

Helps to establish the standard time for an operation or job which has got application in manpower planning, production planning.

Method study – Concept:

Method study enables the industrial engineer to subject each operation to systematic analysis. The main purpose of method study is to eliminate the unnecessary operations and to achieve the best method of performing the operation.

Method study is also called methods engineering or work design. Method engineering is used to describe collection of analysis techniques which focus on improving the effectiveness of men and machines.

According to British Standards Institution (BS 3138), "Method study is the systematic recording and critical examination or existing and proposed ways or doing work as a means or developing and applying easier and more effective methods and reducing cost."

Fundamentally method study involves the breakdown of an operation or procedure into its component elements and their systematic analysis. In carrying out the method study, the right attitude of mind is important.

The method study man should have:

The desire and determination to produce results.

Ability to achieve results.

An understanding of the human factors involved.

Method study scope lies in improving work methods through process and operation analysis, such as:

Manufacturing operations and their sequence.

Workmen.

Materials, tools and gauges.

Layout of physical facilities and work station design.

Movement of men and material handling.

Work environment

Symbols used in method study:

Graphical method of recording was originated by Gilberth, in order to make the presentation of the facts clearly without any ambiguity and to enable to grasp them quickly and clearly. It is useful to use symbols instead of written description.

O OPERATION

□ INSPECTION

→ TRANSPORTATION

D DELAY

∇ STORAGE

Method study is essentially concerned with finding better ways of doing things. It adds value and increases the efficiency by eliminating unnecessary operations, avoidable delays and other forms of waste.

The improvement in efficiency is achieved through:

Improved layout and design of workplace.

Improved and efficient work procedures.

Effective utilization of men, machines and materials.

Improved design or specification of the final product.

Objectives of method study techniques are:

To improve work methods and procedures.

To determine the best sequence of doing work.

To smoothen material flow with minimum of back tracking and to improve layout.

To improve the working conditions and hence to improve labor efficiency.

To reduce monotony in the work.

To improve plant utilization and material utilization.

Elimination of waste and unproductive operations.

To reduce the manufacturing costs through reducing cycle time of operations.

The basic approach to method study consists of the following eight steps.

SELECT the work to be studied and define its boundaries.

RECORD the relevant facts about the job by direct observation and collect such additional data as may be needed from appropriate sources.

EXAMINE the way the job is being performed and challenge its purpose, place sequence and method of performance.

DEVELOP the most practical, economic and effective method, drawing on the contributions of those concerned.

EVALUATE different alternatives to developing a new improved method comparing the cost-effectiveness of the selected new method with the current method with the current method of performance.

DEFINE the new method, as a result, in a clear manner and present it to those concerned, i.e., management, supervisors and workers.

INSTALL the new method as a standard practice and train the persons involved in applying it.

MAINTAIN the new method and introduce control procedures to prevent a drifting back to the previous method of work.

Considerations for selection of method study:

The job should be selected for the method study based upon the following considerations:

Economic aspect: The method study involves cost and time. If sufficient returns are not attained, the whole exercise will go waste. Thus, the money spent should be justified by the savings from it.

The following guidelines can be used for selecting a job:

- ✓ Bottleneck operations which are holding up other production operations.
- ✓ Operations involving excessive labor.
- ✓ Operations producing lot of scrap or defectives.
- ✓ Operations having poor utilization of resources.
- ✓ Backtracking of materials and excessive movement of materials.

Technical Aspects: The method study man should be careful enough to select a job in which he has the technical knowledge and expertise. A person selecting a job in his area of expertise is going to do full justice.

- ✓ Other factors which favor selection in technical aspect are:
- ✓ Job having in consistent quality.
- ✓ Operations generating lot of scraps.
- ✓ Frequent complaints from workers regarding the job.

Human Considerations: Method study means a change as it is going to affect the way in which the job is done presently and is not fully accepted by workman and the union. Human considerations play a vital role in method study. **These are some of the situations where human aspect should be given due importance:**

- ✓ Workers complaining about unnecessary and tiring work.
- ✓ More frequency of accidents.
- ✓ Inconsistent earning.

WORK MEASUREMENT – CONCEPT:

Work measurement is also called by the name 'time study'. Work measurement is absolutely essential for both the planning and control of operations. Without measurement data, we cannot determine the capacity of facilities or it is not possible to quote delivery dates or costs. We are not in a position to determine the rate of production and also labor utilization and efficiency. It may not be possible to introduce incentive schemes and standard costs for budget control.

Objectives of Work Measurement:

The use of work measurement as a basis for incentives is only a small part of its total application. The objectives of work measurement are to provide a sound basis for:

Comparing alternative methods.

Assessing the correct initial manning (manpower requirement planning).

Planning and control.

Realistic costing.

Financial incentive schemes.

Delivery date of goods.

Cost reduction and cost control.

Identifying substandard workers.

Training new employees.

Techniques of Work Measurement:

For the purpose of work measurement, work can be regarded as:

Repetitive work: The type of work in which the main operation or group of operations repeat continuously during the time spent at the job. These apply to work cycles of extremely short duration.

Non-repetitive work: It includes some type of maintenance and construction work, where the work cycle itself is hardly ever repeated identically.

Various techniques of work measurement are:

Time study (stop watch

technique), Synthesis,

Work sampling,

Predetermined motion and time study,

Analytical estimating.

Time study and work sampling involve direct observation and the remaining are data based and analytical in nature.

Time study: A work measurement technique for recording the times and rates of working for the elements of a specified job carried out under specified conditions and for analyzing the data so as to determine the time necessary for carrying out the job at the defined level of performance. In other words measuring the time through stop watch is called time study.

Synthetic data: A work measurement technique for building up the time for a job or pans of the job at a defined level of performance by totaling element times obtained previously from time studies on other jobs containing the elements concerned or from synthetic data.

Work sampling: A technique in which a large number of observations are made over a period of time of one or group of machines, processes or workers. Each observation records what is happening at that instant and the percentage of observations recorded for a particular activity, or delay, is a measure of the percentage of time during which that activities delay occurs.

Predetermined motion time study (PMTS): A work measurement technique whereby times established for basic human motions (classified according to the nature of the motion and conditions under which it is made) are used to build up the time for a job at the defined level of performance. The most commonly used PMTS is known as Methods Time Measurement (MTM).

Analytical estimating: A work measurement technique, being a development of estimating, whereby the time required to carry out elements of a job at a defined level of performance is estimated partly from knowledge and practical experience of the elements concerned and partly from synthetic data.

Time study – concept:

Time study is also called work measurement. It is essential for both planning and control of operations. According to British Standard Institute time study has been defined as "The application of techniques designed to establish the time for a qualified worker to carry out a specified job at a defined level of performance."

Steps in Making Time Study:

Stop watch time is the basic technique for determining accurate time standards. They are economical for repetitive type of work. Steps in taking the time study are:

Select the work to be studied. **Obtain and record** all the information available about the job, the operator and the working conditions likely to affect the time study work. **Breakdown** the operation in to elements. An element is an instinct part of a specified activity composed of one or more fundamental motions selected for convenience of observation and timing. **Measure** the time by means of a stop watch taken by the operator to perform each element of the operation. Either continuous method or snap back method of timing could be used. At the same time, assess the operator's effective speed of work relative to the observer's concept of **'normal'** speed. This is called **performance rating**.

Adjust the observed time by rating factor to obtain normal time for each element; Normal = $\frac{\text{Observed time* Rating}}{100}$

Add the **suitable allowances** to compensate for fatigue, personal needs, and contingencies, to give **standard time** for each element. Compute **allowed time** for the entire job by adding elemental standard times considering frequency of occurrence of each element. Make a detailed job description describing the method for which the standard time is established.

Standard time is the time allowed to an operator to carry out the specified task under specified conditions and defined level of performance. The various allowances are added to the normal time as applicable to get the standard time.

Standard time may be defined as the, amount of time required to complete a unit of work:

Under existing working conditions,

Using the specified method and machinery,

By an operator, able to the work in a proper manner,

and At a standard pace.

Thus basic constituents of standard time are:

Elemental (observed time).

Performance rating to compensate for difference in pace of working. Relaxation allowance.

Interference and contingency allowance.

Steps in time and motion study:

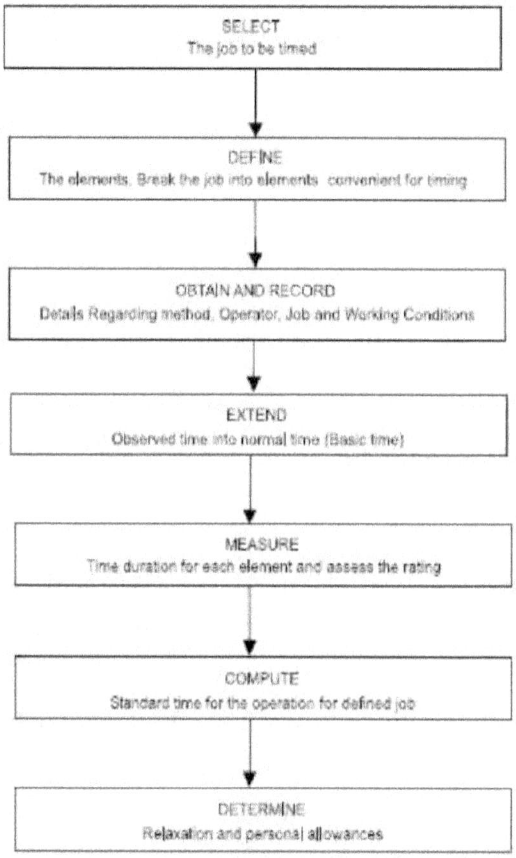

Allowances – concept:

The normal time for an operation does not contain any allowances for the worker. It is impossible to work throughout the day even though the most practicable, effective method has been developed. Even under the best working method situation, the job will still demand the expenditure of human effort and some allowance must therefore be made for recovery from fatigue and for relaxation. Allowances must also be made to enable the worker to attend to his personal needs.

The allowances are categorized as:

Relaxation allowance,

Interference allowance, and

Contingency allowance.

RELAXATION ALLOWANCE

Relaxation allowances are calculated so as to allow the worker to recover from fatigue. Relaxation allowance is a addition to the basic time intended to provide the worker with the opportunity to recover from the physiological and psychological effects of carrying out specified work under specified conditions and to allow attention to personal needs. The amount of allowance will depend on nature of the job.

Relaxation allowances are of two types: fixed allowances and variable allowances. Fixed allowances constitute: Personal needs allowance: It is intended to compensate the operator for the time necessary to leave, the workplace to attend to personal needs like drinking water, smoking, washing hands. Women require longer personal allowance than men. A fair personal allowance is 5% for men, and 7% for women. Allowances for basic fatigue: This allowance is given to compensate for energy expended during working. A common figure considered as allowance is 4% of the basic time. Variable allowance is allowed to an operator who is working under poor environmental conditions that cannot be improved, added stress and strain in performing the job. The variable fatigue allowance is added to the fixed allowance to an operator who is engaged on medium and heavy work and working under abnormal conditions. The amount of variable fatigue allowance varies from organization to organization.

INTERFERENCE ALLOWANCE

It is an allowance of time included into the work content of the job to compensate the operator for the unavoidable loss of production due to simultaneous stoppage of two or more machines being operated by him. This allowance is applicable for machine or process controlled jobs. Interference allowance varies in proportion to number of machines assigned to the operator. The interference of the machine increases the work content.

CONTINGENCY ALLOWANCE

A contingency allowance is a small allowance of time which may be included in a standard time to meet legitimate and expected items of work or delays. The precise measurement of which is uneconomical because of their infrequent or irregular occurrence. This allowance provides for small unavoidable delays as well as for occasional minor extra work.

POLICY ALLOWANCE

Policy allowances are not the genuine part of the time study and should be used with utmost care and only in clearly defined circumstances. The usual reason for making the policy allowance is to line up standard times with requirements of wage agreement between employers and trade unions. The policy allowance is an increment, other than bonus increment, applied to a standard time (or to some constituent part of it, e.g., work content) to provide a satisfactory level of earnings for a specified level of performance under exceptional circumstances. Policy allowances are sometimes made as imperfect functioning of a division or part of a plant.

HUMAN FACTORS IN JOB DESIGN – CONCEPT

"Human factors refer to environmental, organisational and job factors, and human and individual characteristics, which influence behaviour at work in a way which can affect health and safety".

This definition includes three interrelated aspects that must be considered: the job, the individual and the organisation:

The job: including areas such as the nature of the task, workload, the working environment, the design of displays and controls, and the role of procedures. Tasks should be designed in accordance with ergonomic principles to take account of both human limitations and strengths. This includes matching the job to the physical and the mental strengths and limitations of people. Mental aspects would include perceptual, attention and decision making requirements.

The individual: including his/her competence, skills, personality, attitude, and risk perception. Individual characteristics influence behaviour in complex ways. Some characteristics such as personality are fixed; others such as skills and attitudes may be changed or enhanced.

The organisation: including work patterns, the culture of the workplace, resources, communications, leadership and so on. Such factors are often overlooked during the design of jobs but have a significant influence on individual and group behaviour.

In other words, human factors is concerned with what people are being asked to do (the task and its characteristics), who is doing it (the individual and their competence) and where they are working (the organisation and its attributes), all of which are influenced by the wider societal concern, both local and national.

ERGONOMICS – CONCEPT:

Ergonomics is the science which deals with the relationship between man and his working environments. It takes care of factors governing the physical and mental strains. Ergonomics consists of words 'Ergo' (which means work), and 'Nomos' (which means 'Natural Laws'). This can also be termed as 'Human Engineering'.

Ergonomics (or Human Engineering) is defined by I.L.O. (International Labour Organisation) as **"the application of human biological sciences in conjunction with engineering sciences to the worker and his working environment so as to obtain maximum satisfaction for the worker which, at the same time, enhances productivity".**

The task of ergonomics is to develop such conditions for workers, which are necessary to reduce physical workload, to improve working postures, facilitate instrument handling, and thus improves the quality of working life, reduce fatigue, maximize efficiency of production operators and to minimize human errors.

Ergonomics helps to study the effect of working environment on health and safety and in turn on productivity. The workers' interest in the job to a greater extent depends on how comfortable and safe is the workplace.

Objectives of Ergonomics:

Objective of the study of ergonomics is:

To optimize the integration of man and machine in order to increase productivity with accuracy.

To take care of the factors governing the physical and mental strain (i.e. fatigue) so as to get maximum satisfaction for the worker this at the same time enhances the productivity.

Attempts to minimize the risk of injury, illness, accidents and errors without compromising productivity.

To improve the design of machine at the initial design stage or later whenever the existing product or process is modified.

Thus ergonomics helps in:

Developing most comfortable conditions related to climate, lighting, ventilation and noise level;

Reducing the physical work load;

Improving working postures and reducing efforts of certain movements; Making the handling of machine levers and controls easy;

Increasing safety.

Design of Workplace:

Backaches, necaches, and other muscular strains due to bad seating and incorrect working posture are common in industry, where most of the jobs are performed by the operators either in sitting or standing in a fixed posture for a long duration.

The interaction of the operator with the immediate workspace around him is influenced by many factors such as seat design, the working desk and adjacent machine. These factors are responsible for the position and postures of the users and, hence their efficiency.

Factors Considered in the Work Place Design:

1. Purpose of operation.
2. Product design-value analysis.
3. Materials related factors, like.

   Space required, quantity of items, size, etc. for raw, finished and scrap materials.

   Rate of production.

Inspection requirement.

4. Equipment related factors, like

Size.

Utility, service requirements.

Auxiliary equipment.

Number of machines, and space required. Nature of process.

Noise, pollution, vibration, safety hazards etc.

5. Material handling methods.

6. Space related factors; like

Aisles.

Ceiling height required.

Space utilization.

7. Operated related factors:

Sitting or standing.

Level of comfort.

Movement required.

Number of operators.

Supervision requirement.

8. Working Conditions:

Noise, lighting, heating and Ventilation.

Dust, Vibration, window location.

9. Method related factors:

Direction of flow.

Floor levels.

Location of items.

Material movement.

Safety requirement.

Operation sequence.

Tools and material

locations. Production rate.

Auxiliary services.

WORK ENVIRONMENT – CONCEPT:

The work environment refers to the conditions which surround the work place where the worker performs his/her works.

Work factors:

Lighting : lightings should be designed that the employee worker in complete comfort with minimum of eye strain and physical fatigue noise it mean s unwanted sound noise is barrier to workers productivity and wellbeing, every attempt should be made to curb the noise level to its minimum.

Vibration: It should affect performance on target tracking ideally vibrations should be minimized at source thermal conditions human physiology needs a condition of constant temperatures ventilation cool and fresh of air is vital to make an occupied space.

Comfortable: Good house-keeping makes work area look pleasant and more satisfying and reduces fatigue and discomfort.

Selection of colors: Colors influence the workers feelings of warmth and sense of ensure and well being.

Amenities: Amenities should be provided to the workmen without giving them a feeling of generosity examples; canteen should supply good quality food, clean drinking workers safety.

Industrial accidents result from multiplicity of factors, but these have to be traced to their root causes, invariably the fault lies with the management of the organization. The most important function of a healthy and safety programmes is to identify the potential hazards provide safety facilities and equipment's and take prompt remedial action. All employees should be given through guidance in safety methods of work

Health and safety policies:

A health and safety policy is a written statement by an employer stating the company's commitment for the protection of the health and safety of employees and to the public. It is an endorsed commitment by management to its employees regarding their health and safety. A health and safety program contains the health and safety elements of an organization, objectives which make it possible for the company to achieve its goal in the protection of its workers at the workplace.

Section 4(1) of the Occupational Health and Safety Regulations specify the minimum requirements to be contained in a health and safety program. Some of the requirements specified in the regulation may not be applicable to every workplace. However, each employer should carry out their own health and safety risk assessment, in consultation with the occupational health and safety committee, to determine what hazards are present at the workplace. Once the hazards have been identified, controls for exposure to these hazards should be detailed in the health and safety program.

Reasons for health and safety programs or policies in workplace:

There are several reasons why workplaces need a health and safety policy or program, including:

To clearly demonstrate management's full commitment to their employee's health and safety.

To show employees that safety performance and business performance are compatible.

To clearly state the company's safety beliefs, principles, objectives, strategies and processes to build buy-in through all levels of the company.

To clearly outline employer and employee accountability and responsibility for workplace health and safety.

To comply with the Occupational Health and Safety Act; and

To set out safe work practices and procedures to be followed to prevent workplace injuries and illnesses.

Examples of Health and Safety Policies:

Many organizations such as Occupational Health and Safety Branches, Workers' Health, Safety and Compensation Boards, and safety consultants provide good examples of a health and safety policy. For your convenience, the provincial Workers' Health, Safety and Compensation Commission have provided a sample of a health and safety policy below. This is for your viewing and assistance in developing your own health and safety policy. The workers' health and safety representative or the occupational health and safety committees should be consulted when establishing your health and safety policy or programs.

Safety education:

This can include fire drills at a school, proper use of equipment (such as forklifts, scaffolds/ladders, or harnesses) and protective gear (such as hard hats, gloves or safety glasses). It can also include knowing the locations of the fire exits, first aid kit, and fire extinguishers, as well as the overall safety procedures.

Strategies for preventing accidents:

Personal selection

Safety trainings

Environmental factors

Safety campaigns

Safety habits

QUESTION BANK

UNIT – IV

PART – A

S. No	Questions (5 marks)	CO	Level
1	Explain statistical quality control (SQC). List out the various techniques employed.	CO4	L4
2	State the characteristics of control chart.	CO4	L3
3	Give the reason for undertaking time study analysis.	CO4	L3
4	Bring out the benefits of control chart.	CO4	L4
5	"Method study helps in better production goals for employees" Elaborate.	CO4	L4
6	"Inspection is the life blood of quality control" – Discuss.	CO4	L5
7	Discuss how performance rating calculated.	CO4	L4
8	Mention the importance of ergonomics.	CO4	L5
9	Explain pre-determined motion time study.	CO4	L4
10	Explain acceptance sampling.	CO4	L5

PART – B

S. No	Questions (10 marks)	CO	Level
1	"Good working conditions lead to higher productivity" describe.	CO4	L5
2	"Accidents are caused not only due to unsafe acts but also due to unsafe condition-discuss.	CO4	L6
3	Elaborate basic consideration for selection of method study.	CO4	L5
4	Summarize the basic procedure of method study.	CO4	L6
5	Analyze the techniques of work measurements.	CO4	L5
6	Elucidate categories of allowances.	CO4	L5

PART – C

CASE STUDY – (20 marks)

<table>
<tr><th>S. No</th><th>Questions (20 marks)</th><th>CO</th><th>Level</th></tr>
<tr><td>1</td><td>The Assam Gas Cracker Project conceived as part of the Assam Accord signed in 1985 is yet to see the light of the day. It has been plagued by a host of problems starting from location to economic viability. Originally planned at Tengakhat, it was later shifted to a place called Lepetkata. The project is now being implemented by GAIL (a Government of India enterprise) as the lead promoter (70% share) with another public enterprise OIL (20% share) and the Government of Assam as minor Himadri Barman, Centre for Management Studies, Dibrugarh University, Dibrugarh 786 004 (Assam) Downloaded from http://himadri.cmsdu.org 2 partners. GAIL had to be brought in after India's largest private sector enterprise Reliance Industries backed out of the project saying that it was economically unviable. The land acquisition for the project (as of mid 2008) is yet to be completed and there is still a lot of uncertainty regarding the availability of raw materials for production. In the meantime, the project cost has spiraled many times over to INR 50 billion, which is likely to go up further.
(a) Discuss the importance of Project Management in the light of the above situation.
(b) As a project manager employed with GAIL, what would be your line of action to see to it that the project is not delayed any further?
(c) Why do projects suffer from time and cost overruns?</td><td>CO4</td><td>L6</td></tr>
<tr><td>2</td><td>Ten samples (each of size 100) of a component were inspected. The results of the inspection are given below.
<table>
<tr><td>Sample No:</td><td>1</td><td>2</td><td>3</td><td>4</td><td>5</td><td>6</td><td>7</td><td>8</td><td>9</td><td>10</td></tr>
<tr><td>No. Of defects</td><td>2</td><td>0</td><td>4</td><td>3</td><td>1</td><td>2</td><td>3</td><td>1</td><td>1</td><td>2</td></tr>
</table>
Draw the relevant control chart taking 3 sigma limits.</td><td>CO4</td><td>L6</td></tr>
</table>

TEXT / REFERENCE BOOKS

R. Paneerselvam, Production and Operations Management, PHI Learning Private Ltd, 3rd edition, 2012.

R. B. Kanna, Production and Operations Management, PHI Learning Private Ltd, 2nd edition, 2015.

P. Saravanavel & S.Sumathi, Production and Materials Management, Margham Publications, 2nd edition, 2012.

S. N. Chary, Production and Operations Management, Tata McGraw Hill Education Private Limited, 4th edition, 2009.

S. P. Singh, Production and Operations Management, Vikas Publishing House Private Limited, 1st edition, 2014.

V. MATERIALS MANAGEMENT AND INVENTORY CONTROL

Aggregate Planning – Nature, Strategies, Methods - Material Requirement Planning (MRP-I) – Manufacturing Resource Planning (MRP-II) – Inventory – Types of Inventory - Deterministic demand model – EOQ - Continuous and Periodic review Inventory models- Selective Inventory Control – ABC, VED, FSN Techniques.

AGGREGATE PLANNING – CONCEPT:

An organization can finalize its business plans on the recommendation of demand forecast. Once business plans are ready, an organization can do backward working from the final sales unit to raw materials required. Thus annual and quarterly plans are broken down into labor, raw material, working capital, etc. requirements over a medium-range period (6 months to 18 months). This process of working out production requirements for a medium range is called aggregate planning.

Factors Affecting Aggregate Planning:

Aggregate planning is an operational activity critical to the organization as it looks to balance long-term strategic planning with short term production success. Following factors are critical before an aggregate planning process can actually start;

- A complete information is required about available production facility and raw materials.
- A solid demand forecast covering the medium-range period
- Financial planning surrounding the production cost which includes raw material, labor, inventory planning, etc.
- Organization policy around labor management, quality management, etc.

For aggregate planning to be a success, following inputs are required;

- An aggregate demand forecast for the relevant period.
- Evaluation of all the available means to manage capacity planning like sub-contracting, outsourcing, etc.
- Existing operational status of workforce (number, skill set, etc.), inventory level and production efficiency
- Aggregate planning will ensure that organization can plan for workforce level, inventory level and production rate in line with its strategic goal and objective.

Aggregate planning as an Operational Tool

Aggregate planning helps achieve balance between operation goal, financial goal and overall strategic objective of the organization. It serves as a platform to manage capacity and demand planning. In a scenario where demand is not matching the capacity, an organization can try to balance both by pricing, promotion, order management and new demand creation. In scenario where capacity is not matching demand, an organization can try to balance the both by various alternatives such as.

Laying off/hiring excess/inadequate workforce until demand decrease/increase.

Including overtime as part of scheduling there by creating additional capacity.

Hiring a temporary workforce for a fix period or outsourcing activity to a sub-contractor.

Importance of Aggregate Planning:

Aggregate planning plays an important part in achieving long-term objectives of the organization. Achieving financial goals by reducing overall variable cost and improving the bottom line

Maximum utilization of the available production facility

Provide customer delight by matching demand and reducing wait time for customers Reduce investment in inventory stocking

Able to meet scheduling goals there by creating a happy and satisfied work force

Aggregate Planning Strategies:

There are three types of aggregate planning strategies available for organization to choose from. They are as follows.

Level Strategy: As the name suggests, level strategy looks to maintain a steady production rate and workforce level. In this strategy, organization requires a robust forecast demand as to increase or decrease production in anticipation of lower or higher customer demand. Advantage of level strategy is steady workforce. Disadvantage of level strategy is high inventory and increase back logs.

Chase Strategy: As the name suggests, chase strategy looks to dynamically match demand with production. Advantage of chase strategy is lower inventory levels and back logs. Disadvantage is lower productivity, quality and depressed work force.

Hybrid Strategy: As the name indicates, the Hybrid strategy is an integration of both level and chase strategies to get a better result. It maintains a sufficient balance between stock level, recruiting, termination and production rate. In the hybrid strategy of aggregate planning, the organizations build up inventory before rising demands. It uses backorders to level with high peak periods. It can easily cover short-term peaks by hiring workers temporarily or by subcontracting production. Hiring, lay-off and reassigning workers is a normal part of the hybrid strategy.

Techniques of Aggregate Planning:

Various techniques are used to perform the task of aggregate planning. Usually, there are two categories: **Informal trial-and-error techniques and mathematical techniques.** In practice, informal techniques are more commonly used. However, a substantial amount of research has been done to mathematical techniques, but still, they are not as extensively used, they often serve as a basis for comparing the effectiveness of alternative techniques for aggregate planning.

There are several steps in general procedure for aggregate planning:

Determine demand for each period.

Determine capacities (regular time, overtime, subcontracting) for each period.

Identify company or departmental policies that are pertinent (e.g., maintain a safety stock of 5 percent of demand, maintain a reasonably stable workforce).

Determine unit costs for regular time, overtime, subcontracting, holding inventories, back orders, layoffs, and other relevant costs.

Develop alternative plans and compute the cost for each activity.

If satisfactory plans emerge, select the one that best satisfies objectives. Otherwise, return to step 5.

The advantages associated with aggregate planning include-

It helps the organization in dealing with production facilities in a lean manner.

The process helps to develop effective strategic plan as well as relationships with distributors and suppliers. It also assists in making developing accurate market research

The planning helps in the optimization of inventory.

An essential advantage is that it serves as a useful tool for making viable forecasts about product demand.

It helps to adjust capacity so that it can meet demands.

The aggregate planning process helps to calculate capacity, for instance, how many units can be produced daily or in a week or a month.

It helps the organization to identify the best options so that it can meet the demands easily.

It assists in knowing about the inefficiencies that exist within the organization

It helps to determine resources for instance amount of raw materials on hand, availability of total machine hours and the total number of workers along with products in progress, packaging materials, and tools required for manufacturing finished goods.

An advantage of the planning process is that it helps to project demand and figure out the units in need for the short-term by factoring in advertising campaigns, special pricing, and promotions.

It encourages the optimized utilization of space.

MATERIAL REQUIREMENTS PLANNING – CONCEPT:

Material Requirements Planning (or MRP) is a method that is used for the purpose of calculating the components and the materials, which in turn will be needed for the sake of making a product. Ideally, one can say that it has three broad, main steps. They would be as follows:

Taking note of the materials' inventory and the components which are readily available on hand.

To identify the additional ones this would be required for the sake of seamless working of the manufacturing and production processes.

Scheduling of production or the purchase of raw materials and components, as there might be a need to ensure that there are no unwanted delays in the production pipeline.

"Materials Requirement Planning (MRP) is a technique for determining the quantity and timing for the acquisition of dependent demand items needed to satisfy master production schedule requirements."

Objectives of Material Requirement Planning:

Inventory reduction: MRP determines how many components are required when they are required in order to meet the master schedule. It helps to procure the materials/ components as and when needed and thus avoid excessive build up of inventory.

Reduction in the manufacturing and delivery lead times: MRP identifies materials and component quantities, timings when they are needed, availabilities and procurements and actions required to meet delivery deadlines. MRP helps to avoid delays in production and priorities production activities by putting due dates on customer job order.

Realistic delivery commitments: By using MRP, production can give marketing timely information about likely delivery times to prospective customers.

Increased efficiency: MRP provides a close coordination among various work centers and hence help to achieve uninterrupted flow of materials through the production line. This increases the efficiency of production system.

MRP Process:

The inputs to the MRP system are: (1) A master production schedule, (2) An inventory status file and (3) Bill of materials (BOM). Using these three information sources, the MRP processing logic (computer programme) provides three kinds of information (output) for each product component: order release requirements, order rescheduling and planned orders.

MRP system

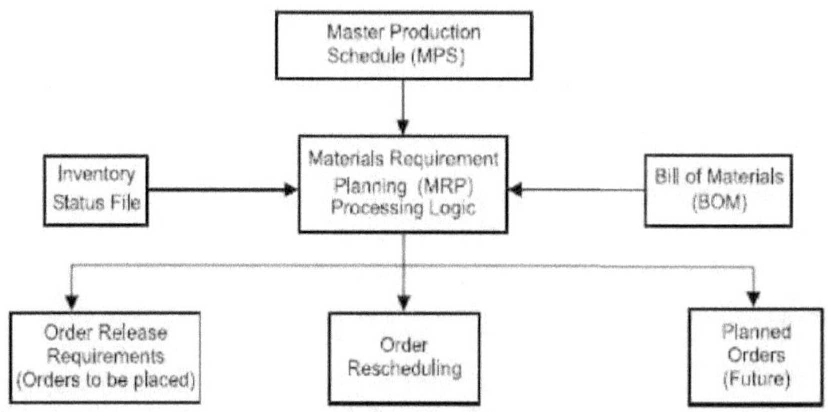

MASTER PRODUCTION SCHEDULE (MPS): MPS is a series of time phased quantities for each item that a company produces, indicating how many are to be produced and when. MPS is initially developed from firm customer orders or from forecasts of demand before

MRP system begins to operate. The MRP system whatever the master schedule demands and translates MPS end items into specific component requirements. Many systems make a simulated trial run to determine whether the proposed master can be satisfied.

INVENTORY STATUS FILE: Every inventory item being planned must have an inventory status file which gives complete and up to date information on the on-hand quantities, gross requirements, scheduled receipts and planned order releases for an item. It also includes planning information such as lot sizes, lead times, safety stock levels and scrap allowances.

BILL OF MATERIALS (BOM): BOM identifies how each end product is manufactured, specifying all subcomponents items, their sequence of build up, their quantity in each finished unit and the work centers performing the buildup sequence. This information is obtained from product design documents, workflow analysis and other standard manufacturing information.

What is Manufacturing Resource Planning?

Manufacturing Resource Planning (MRP II) is an integrated information system used by businesses. Manufacturing Resource Planning (MRP II) evolved from early Materials Requirement Planning (MRP) systems by including the integration of additional data, such as employee and financial needs.

The system is designed to centralize, integrate, and process information for effective decision making in scheduling, design engineering, inventory management, and cost control in manufacturing.

Both MRP and MRP II are seen as predecessors to Enterprise resource planning (ERP), which is a process whereby a company, often a manufacturer, manages and integrates the important parts of its business.

An ERP management information system integrates areas such as planning, purchasing, inventory, sales, marketing, finance, and human resources. ERP is most frequently used in the context of software, with many large applications having been developed to help companies implement ERP.

Understanding Manufacturing Resource Planning (MRP II)

MRP II is a computer-based system that can create detailed production schedules using real-time data to coordinate the arrival of component materials with machine and labor availability. MRP II is used widely by itself, but it's also used as a module of more extensive enterprise resource planning (ERP) systems.

MRP II is an extension of the original materials requirements planning (MRP I) system.

Materials requirements planning (MRP) is one of the first software-based integrated information systems designed to improve productivity for businesses.

A materials requirements planning information system is a sales forecast-based system used to schedule raw material deliveries and quantities, given assumptions of machine and labor units required to fulfill a sales forecast.

By the 1980s, manufacturers realized they needed software that could also tie into their accounting systems and forecast inventory requirements. MRP II was provided as a solution, which included this functionality in addition to all the capabilities offered by MRP I.

MRP I vs. MRP II:

For all intents and purposes, MRP II has effectively replaced MRP I software. Most MRP II systems deliver all of the functionality of an MRP system. But in addition to offering master production scheduling, bill of materials (BOM), and inventory tracking, MRP II provides functionality within logistics, marketing, and general finance.

For example, MRP II is able to account for variables that MRP is not including machine and personnel capacity providing a more realistic and holistic representation of a company's operating capabilities. Many MRP II solutions also offer simulation features that allow operators to enter variables and see the downstream effect. Because of its ability to provide feedback on a given operation, MRP II is sometimes referred to as a closed-loop system.

MRP I included the following three major functionalities:

master production scheduling

bill of materials

inventory tracking

MRP II includes those three, plus the following:

machine capacity scheduling

demand forecasting

quality assurance

general accounting

MRP II systems are still in wide use by manufacturing companies today and can either be found as stand-alone solutions or as part of an enterprise resource planning (ERP) system. Enterprise Resources Planning (ERP) software systems are regarded as the successors of MRP II software. ERP suites include applications well outside the scope of manufacturing. These can include everything from human resources and customer relationship management to enterprise asset management.

Meaning of Inventory:

Inventory generally refers to the materials in stock. It is also called the idle resource of an enterprise. Inventories represent those items which are either stocked for sale or they are in the process of manufacturing or they are in the form of materials, which are yet to be utilized. The interval between receiving the purchased parts and transforming them into final products varies from industries to industries depending upon the cycle time of manufacture. It is, therefore, necessary to hold inventories of various kinds to act as a buffer between supply and demand for efficient operation of the system. Thus, an effective control on inventory is a must for smooth and efficient running of the production cycle with least interruptions.

Reasons for Keeping Inventories:

To stabilize production: The demand for an item fluctuates because of the number of factors, e.g., seasonality, production schedule etc. The inventories (raw materials and components) should be made available to the production as per the demand failing which results in stock out and the production stoppage takes place for want of materials. Hence, the inventory is kept to take care of this fluctuation so that the production is smooth.

To take advantage of price discounts: Usually the manufacturers offer discount for bulk buying and to gain this price advantage the materials are bought in bulk even though it is not required immediately. Thus, inventory is maintained to gain economy in purchasing.

To meet the demand during the replenishment period: The lead time for procurement of materials depends upon many factors like location of the source, demand supply condition, etc. So inventory is maintained to meet the demand during the procurement (replenishment) period.

To prevent loss of orders (sales): In this competitive scenario, one has to meet the delivery schedules at 100 per cent service level, means they cannot afford to miss the delivery schedule which may result in loss of sales. To avoid the organizations have to maintain inventory.

To keep pace with changing market conditions: The organizations have to anticipate the changing market sentiments and they have to stock materials in anticipation of non- availability of materials or sudden increase in prices. Sometimes the organizations have to stock materials due to other reasons like suppliers minimum quantity condition, seasonal availability of materials or sudden increase in prices.

Meaning of Inventory Control:

Inventory control is a planned approach of determining what to order, when to order and how much to order and how much to stock so that costs associated with buying and storing are optimal without interrupting production and sales. Inventory control basically deals with two problems:

(i) When should an order be placed? (Order level), and

(ii) How much should be ordered? (Order quantity).

These questions are answered by the use of inventory models. The scientific inventory control system strikes the balance between the loss due to non-availability of an item and cost of carrying the stock of an item. Scientific inventory control aims at maintaining optimum level of stock of goods required by the company at minimum cost to the company.

Objectives of Inventory Control:

To ensure adequate supply of products to customer and avoid shortages as far as possible.

To make sure that the financial investment in inventories is minimum (i.e., to see that the working capital is blocked to the minimum possible extent).

Efficient purchasing, storing, consumption and accounting for materials is an important objective.

To maintain timely record of inventories of all the items and to maintain the stock within the desired limits.

To ensure timely action for replenishment.

To provide a reserve stock for variations in lead times of delivery of materials.

To provide a scientific base for both short-term and long-term planning of materials.

Benefits of Inventory Control:

It is an established fact that through the practice of scientific inventory control, following are the benefits of inventory control:

Improvement in customer's relationship because of the timely delivery of goods and service.

Smooth and uninterrupted production and, hence, no stock out.

Efficient utilization of working capital. Helps in minimizing loss due to deterioration, obsolescence damage and pilferage.

Economy in purchasing.

Eliminates the possibility of duplicate ordering.

Techniques of Inventory Control:

In any organization, depending on the type of business, inventory is maintained. When the number of items in inventory is large and then large amount of money is needed to create such inventory, it becomes the concern of the management to have a proper control over its ordering, procurement, maintenance and consumption. The most widely used method of inventory control is known as ABC analysis. In this technique, the total inventory is categorized into three sub- heads and then proper exercise is exercised for each sub-heads.

ABC analysis: In this analysis, the classification of existing inventory is based on annual consumption and the annual value of the items. Hence we obtain the quantity of inventory item consumed during the year and multiply it by unit cost to obtain annual usage cost. The items are then arranged in the descending order of such annual usage cost. The analysis is carried out by drawing a graph based on the cumulative number of items and cumulative usage of consumption cost.

Category	*Percentage of items*	*Percentage of annual consumption value*
A	10–20	70–80
B	20–30	10–25
C	60–70	5–15

The classification of ABC analysis is shown by the graph

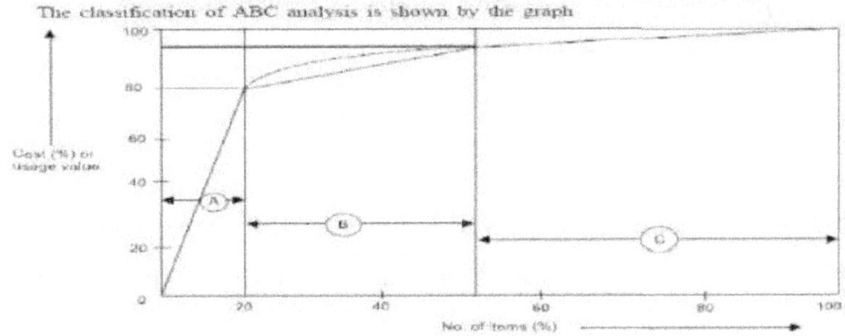

Once ABC classification has been achieved, the policy control can be formulated as follows:
A-Item: Very tight control, the items being of high value. The control need be exercised at higher level of authority.
B-Item: Moderate control, the items being of moderate value. The control need be exercised at middle level of authority.
C-Item: The items being of low value, the control can be exercised at gross root level of authority, i.e., by respective user department managers.

HML analysis: In this analysis, the classification of existing inventory is based on unit price of the items. They are classified as high price, medium price and low cost items.

VED analysis: In this analysis, the classification of existing inventory is based on criticality of the items. They are classified as vital, essential and desirable items. It is mainly used in spare parts inventory.

FSN analysis: In this analysis, the classification of existing inventory is based consumption of the items. They are classified as fast moving, slow moving and non-moving items.

SDE analysis: In this analysis, the classification of existing inventory is based on the items.

GOLF analysis: In this analysis, the classification of existing inventory is based sources of the items. They are classified as Government supply, ordinarily available, local availability and foreign source of supply items.

SOS analysis: In this analysis, the classification of existing inventory is based nature of supply of items. They are classified as seasonal and off-seasonal items.

For effective inventory control, combination of the techniques of ABC with VED or ABC with HML or VED with HML analysis is practically used.

Inventory Model:

ECONOMIC ORDER QUANTITY (EOQ):

Inventory models deal with idle resources like men, machines, money and materials. These models are concerned with two decisions: how much to order (purchase or produce) and when to order so as to minimize the total cost.

For the first decision how much to order, there are two basic costs are considered namely, inventory carrying costs and the ordering or acquisition costs. As the quantity ordered is increased, the inventory carrying cost increases while the ordering cost decreases. The 'order quantity' means the quantity produced or procured during one production cycle. Economic order quantity is calculated by balancing the two costs. Economic Order Quantity (EOQ) is that size of order which minimizes total costs of carrying and cost of ordering.

i.e., Minimum Total Cost occurs when Inventory Carrying Cost = Ordering Cost Economic order quantity can be determined by two methods:

Tabulation method.

Algebraic method

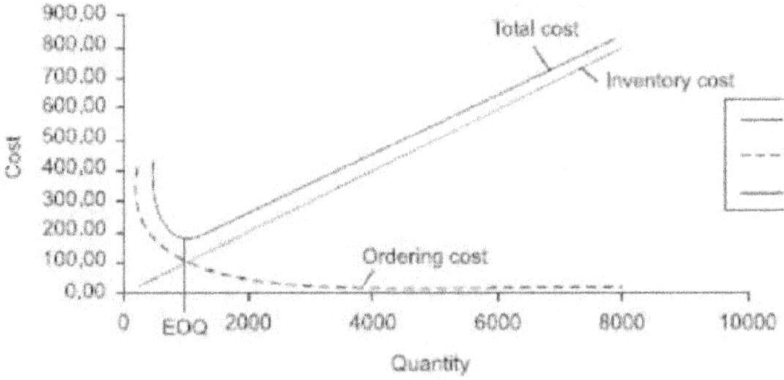

Determination of EOQ by Tabulation (Trial & Error) Method:

This method involves the following steps:

Select the number of possible lot sizes to purchase.

Determine average inventory carrying cost for the lot purchased.

Determine the total ordering cost for the orders placed.

Determine the total cost for each lot size chosen which is the summation of inventory carrying cost and ordering cost.

Select the ordering quantity, which minimizes the total cost

The XYZ Ltd. carries a wide assortment of items for its customers. One of its popular items has annual demand of 8000 units. Ordering cost per order is found to be Rs. 12.5. The carrying cost of average inventory is 20% per year and the cost per unit is Re. 1.00. Determine the optimal economic quantity and make your recommendations.

SOLUTION:

No. of orders/ year (1)	*Lot size (2)*	*Average inventory (3)*	*Carrying cost (4)*	*Ordering cost (5)*	*Total cost/ year (6) =(4) + (5)*
1	8000	4000	800.00	12.5	812.50
2	4000	2000	400.00	25	425.00
4	2000	1000	200.00	50	250.00
8	1000	500	100.00	100	200.00
12	666.667	333.333	66.67	150	216.67
16	500	250	50.00	200	250.00

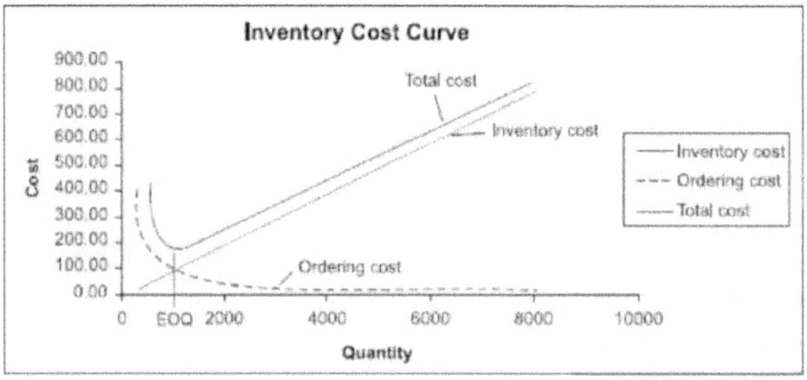

The table and the graph indicates that an order size of 1000 units will gives the lowest total cost among the different alternatives. It also shows that minimum total cost occurs when carrying cost is equal to ordering cost.

Determination of EOQ by Analytical Method:

In order to derive an economic lot size formula following assumptions are made:

Demand is known and uniform.

Let D denotes the total number of units purchase/produced and Q denotes the lot size in each production run.

Shortages are not permitted, i.e., as soon as the level of the inventory reaches zero, the inventory is replenished.

Production or supply of commodity is instantaneous.

Lead-time is zero. Set-up cost per production run or procurement cost is C3.Inventory carrying cost is C1 = CI, where C is the unit cost and I is called inventory carrying cost

The most economic point in terms of total inventory cost exists where,

Inventory carrying cost = Annual ordering cost (set-up cost)

Average inventory = 1/2 (maximum level + minimum level)

= (Q + 0)/2 = Q/2

Total inventory carrying cost = Average inventory × Inventory carrying cost per unit

i.e., Total inventory carrying cost $= Q/2 \times C_1 = QC_1/2$ ...(1)

Total annual ordering costs = Number of orders per year × Ordering cost per order

i.e., Total annual ordering costs $= (D/Q) \times C_3 = (D/Q)C_3$ ...(2)

Now, summing up the total inventory cost and the total ordering cost, we get the total inventory cost C(Q).

i.e., Total cost of production run = Total inventory carrying cost + Total annual ordering costs

$$C(Q) = QC_1/2 + (D/Q)C_3 \quad \text{(cost equation)} \qquad ...(3)$$

But, the total cost is minimum when the inventory carrying costs becomes equal to the total annual ordering costs. Therefore,

$$QC_1/2 = (D/Q)C_3$$

or $$QC_1 = (2D/Q)C_3 \quad \text{or} \quad Q^2 = 2C_3D/C_1$$

or $$Q = \sqrt{\frac{2C_3D}{C_1}}$$

i.e., Optimal quantity (EOQ), $Q_0 = \sqrt{\frac{2C_3D}{C_1}}$ ...(4)

expressed as a percentage of the value of the average inventory.

$$\text{Optimum number of orders, } (N_0) = \frac{D}{Q_0} \qquad ...(5)$$

$$\text{Optimum order interval, } (t_0) = \frac{365}{N_0} \text{ in days} = \frac{1}{N_0} \text{ in years or } (t_0) = \frac{Q_0}{D} \qquad ...(6)$$

$$\text{Average yearly cost (TC)} = \sqrt{2C_3DC_1} \qquad ...(7)$$

ILLUSTRATION 4: *An oil engine manufacturer purchases lubricants at the rate of Rs. 42 per piece from a vendor. The requirements of these lubricants are 1800 per year. What should be the ordering quantity per order, if the cost per placement of an order is Rs. 16 and inventory carrying charges per rupee per year is 20 paise.*

SOLUTION: Given data are:

Number of lubricants to be purchased, D = 1800 per year

Procurement cost, C_3 = Rs. 16 per order

Inventory carrying cost, $CI = C_1$ = Rs. 42 × Re. 0.20 = Rs. 8.40 per year

Then, optimal quantity (EOQ), $Q_0 = \sqrt{\frac{2C_3D}{C_1}}$

$$Q_0 = \sqrt{\frac{2 \times 16 \times 1800}{8.4}} = 82.8 \text{ or } 83 \text{ lubricants (approx).}$$

QUESTION BANK UNIT – V

PART – A

S. No	Questions (5 marks)	CO	Level
1	List down the objectives inventory control.	CO5	L3
2	Mention the importance of inventory control.	CO5	L3
3	Define aggregate planning and its objectives.	CO5	L5
4	Explain the benefits of inventory control.	CO5	L4
5	Bring out the meaning of BOM.	CO5	L3
6	Give the conditions required for MRP.	CO5	L3
7	Expand FSN and bring out its importance.	CO5	L3
8	Indicate the significance of inventories management.	CO5	L5
9	Discuss an MRP system which helps management to monitor the performance of the inventory system.	CO5	L5
10	Explain the different selective inventory control techniques.	CO5	L5

PART – B

S. No	Questions (10 marks)	CO	Level
1	Summarize MRP I and MRP II system in detail.	CO5	L5
2	Enumerate techniques of inventory control.	CO5	L6
3	"Ideal inventory management brings the cost of the product down" Explain.	CO5	L5
4	Compare and contrast any ABC & VED inventory techniques.	CO5	L6
5	Elucidate ABC analysis in operation management	CO5	L5
6	Generalize the lead time in inventory really important how can mismanaging lead time cost a company money	CO5	L6

PART – C

CASE STUDY – (20 marks)

S. No	Questions (20 marks)	CO	Level
1	A local distributor for a national tyre company expects to sell approximately 9600 steel belted radial tyres of a certain size and tread design next year. Annual carrying cost is Rs.160 per tyre and ordering cost is Rs.750. The distributor operates 288 days a year. What is the Economic Order Quantity (EOQ)?	CO5	L6
2	Ashok Leyland has the good market share of trucks and passenger transport vehicles across India. The production capacity of AL is overall 8000 trucks per month in south India and North and West part of India manufactures around 20000 trucks. Market demand surges due to recession and the requirement has surged to 8000 trucks overall India. But the management believes the market will boom and they need to maintain around 40000 trucks for the 2 months despite of surge in the sales. The demand is estimated to surge @ 8% and during the boom it will be around 26% increase. The alternate for the company is to hold the inventory for 2 months and it will be ready to observe the inventory holding cost which accounts 15% of their working capital. Otherwise management could not able to supply the trucks in time to the customers. Now the management has to take the decision on the following factors. 1. How much inventory they should keep during recession and boom period? 2. What should be production capacity to be maintained in both south and north factories? Please note demand overall India.	CO5	L6

TEXT / REFERENCE BOOKS

R. Paneerselvam, Production and Operations Management, PHI Learning Private Ltd, 3rd edition, 2012.

R. B. Kanna, Production and Operations Management, PHI Learning Private Ltd, 2nd edition, 2015.

P. Saravanavel & S.Sumathi, Production and Materials Management, Margham Publications, 2nd edition, 2012.

S. N. Chary, Production and Operations Management, Tata McGraw Hill Education Private Limited, 4th edition, 2009.

S. P. Singh, Production and Operations Management, Vikas Publishing House Private Limited, 1st edition, 2014.

Problem

Develop a network diagram for the project specified below:

Activity	Immediate Predecessor Activity
A	-
B	A
C, D	B
E	C
F	D
G	E, F

Solution

Activity A has no predecessor activity. i.e., It is the first activity. Let us suppose that activity A takes the project from event 1 to event 2. Then we have the following representation for A:

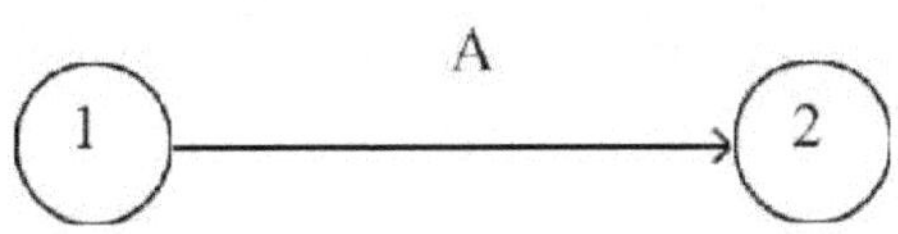

For activity B, the predecessor activity is A. Let us suppose that B joins nodes 2 and 3. Thus we get

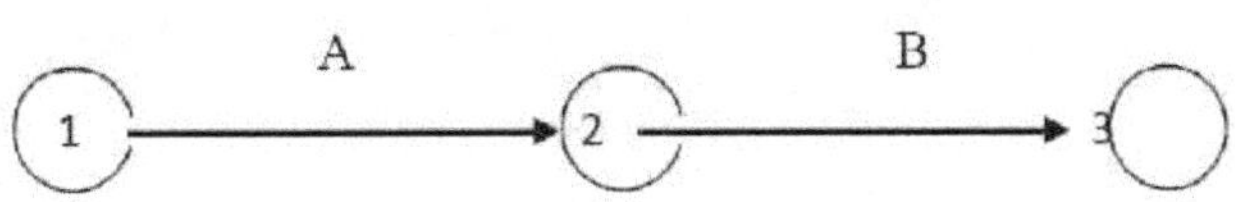

Activities C and D have B as the predecessor activity. Therefore we obtain the following:

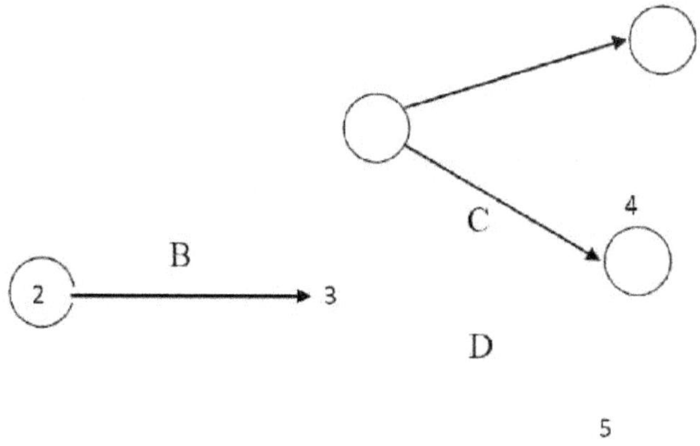

Activity E has D as the predecessor activity. So we get

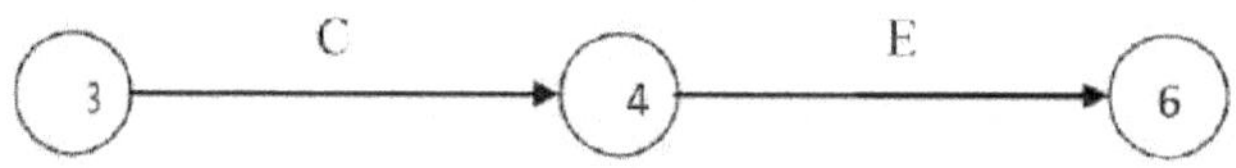

Activity F has D as the predecessor activity. So we get

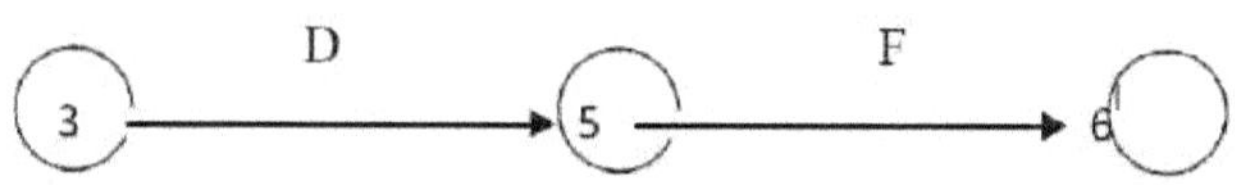

Activity G has E and F as predecessor activities. This is possible only if nodes 6 and 6l are one and the same. So, rename node 6l as node 6. Then we get

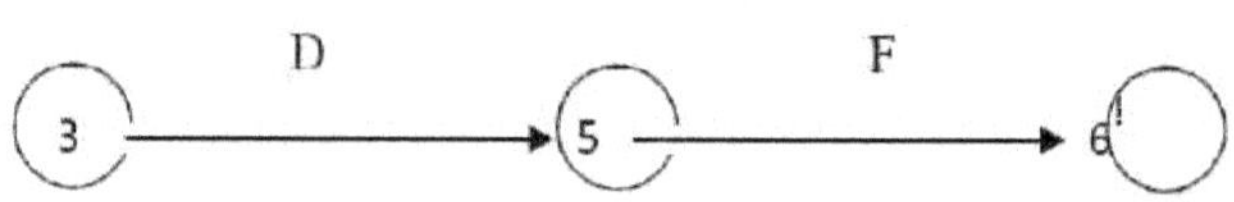

and

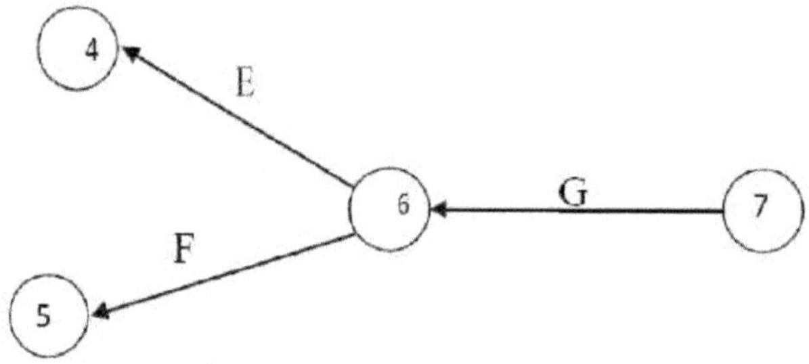

G is the last activity.

Putting all the pieces together, we obtain the following diagram the project network:

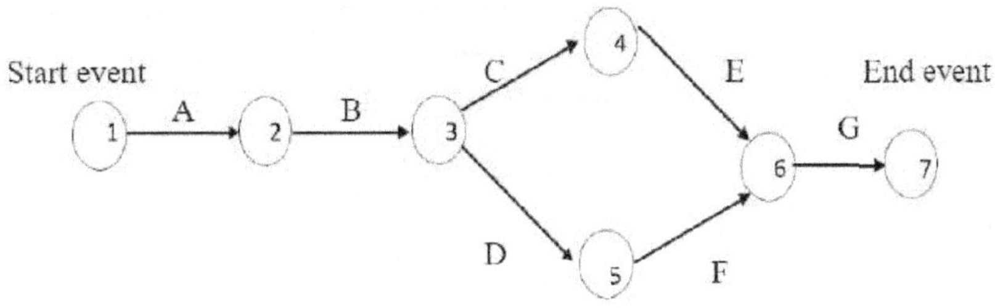

The diagram is validated by referring to the given data.

Note: An important point may be observed for the above diagram. Consider the following parts in the diagram

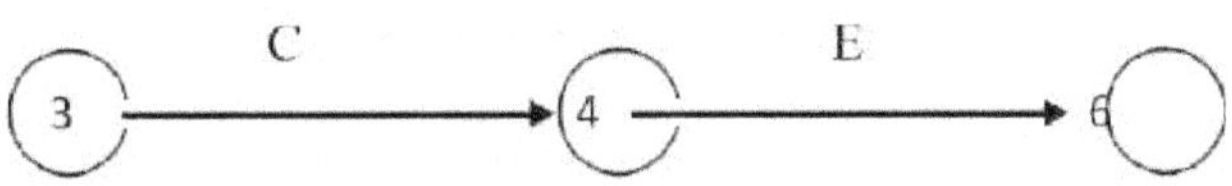

and

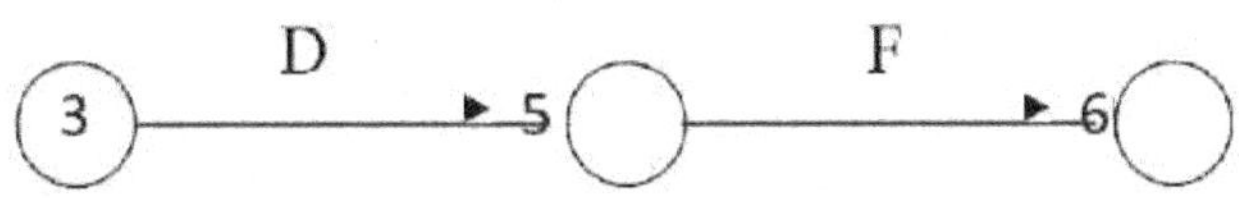

We took nodes 6 and 6l as one and the same. Instead, we can retain them as different nodes. Then, in order to provide connectivity to the network, we join nodes 6l and 6 by a dummy activity. Then we arrive at the following diagram for the project network:

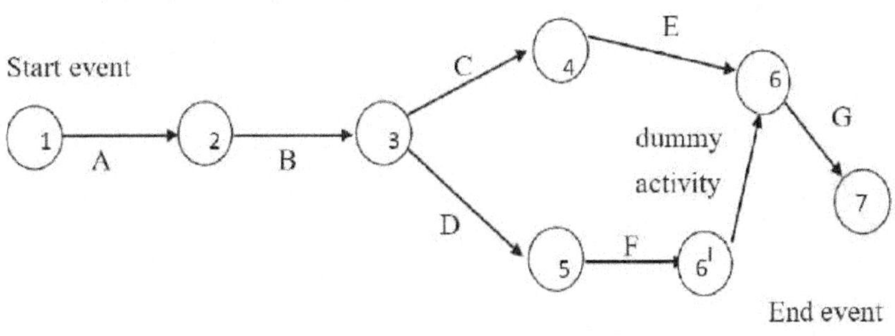

Questions

1. Explain the terms: event, predecessor event, successor event, activity, dummy activity, network.
2. Construct the network diagram for the following project:

Activity	Immediate Predecessor Activity
A	-
B	-
C	A
D	B
E	A
F	C, D
G	E
H	E
I	F, G
J	H, I

****

Critical Path Method (CPM)

Lesson Outline

- The concepts of critical path and critical activities
- Location of the critical path
- Evaluation of the project completion time

Learning Objectives

After reading this lesson you should be able to

- understand the definitions of critical path and critical activities
- identify critical path and critical activities
- determine the project completion time

Introduction

The critical path method (CPM) aims at the determination of the time to complete a project and the important activities on which a manager shall focus attention.

Assumption For Cpm

In CPM, it is assumed that precise time estimate is available for each activity.

Project Completion Time

From the start event to the end event, the time required to complete all the activities of the project in the specified sequence is known as the project completion time.

Path In A Project

A continuous sequence, consisting of nodes and activities alternatively, beginning with the start event and stopping at the end event of a network is called a path in the network.

Critical Path And Crtical Activities

Consider all the paths in a project, beginning with the start event and stopping at the end event. For each path, calculate the time of execution, by adding the time for the individual activities in that path.

The path with the largest time is called the critical path and the activities along this path are called the critical activities or bottleneck activities. The activities are called critical because they cannot be delayed. However, a non-critical activity may be delayed to a certain extent. Any delay in a critical activity will delay the completion of the whole project. However, a certain permissible delay in a non –critical activity will not delay the completion of the whole project. It shall be noted that delay in a non-critical activity beyond a limit would certainly delay the completion the whole project. Sometimes, there may be several critical paths for a project. A project manager shall pay special attention to critical activities.

Problem 1

The following details are available regarding a project:

Activity	Predecessor Activity	Duration (Weeks)
A	-	3
B	A	5
C	A	7
D	B	10
E	C	5
F	D,E	4

Determine the critical path, the critical activities and the project completion time.

Solution

First let us construct the network diagram for the given project. We mark the time estimates along the arrows representing the activities. We obtain the following diagram:

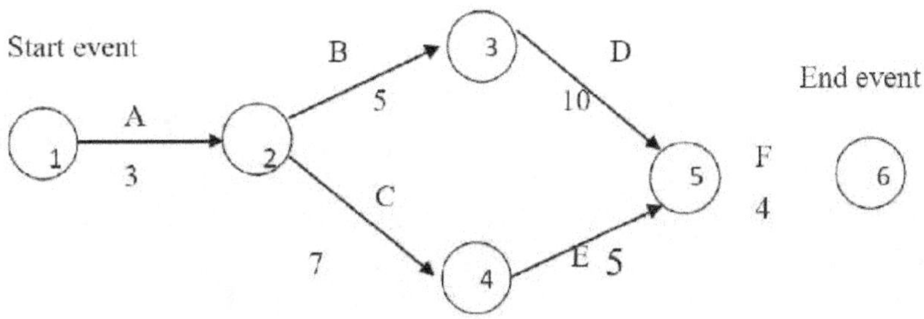

Consider the paths, beginning with the start node and stopping with the end node. There are two such paths for the given project. They are as follows:

Path I

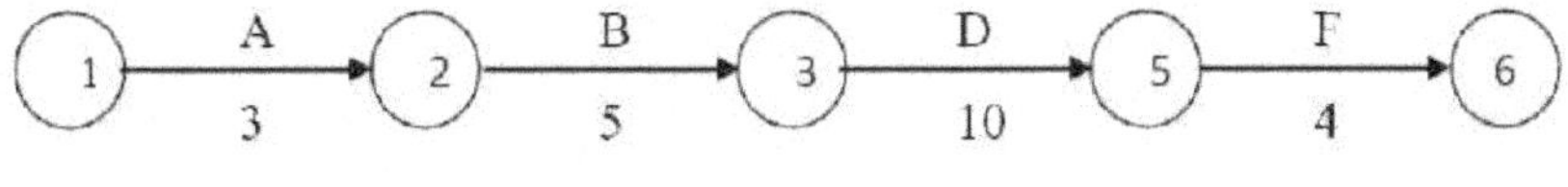

with a time of 3 + 5 + 10 + 4 = 22 weeks.

Path II

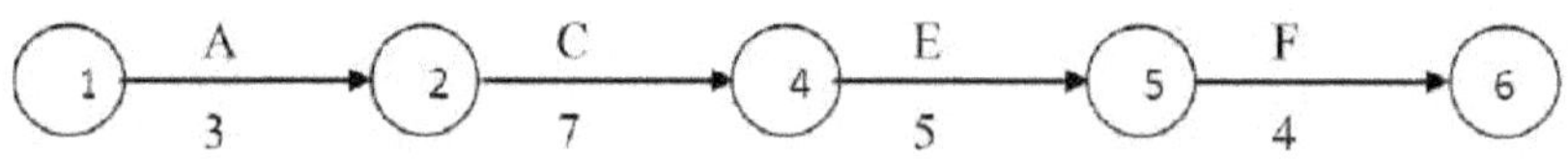

with a time of 3 + 7 + 5 + 4 = 19 weeks.

Compare the times for the two paths. Maximum of {22,19} = 22. We see that path I has the maximum time of 22 weeks. Therefore, path I is the critical path. The critical activities are A, B, D and F. The project completion time is 22 weeks.

We notice that C and E are non- critical activities.

Time for path I - Time for path II = 22- 19 = 3 weeks.

Therefore, together the non- critical activities can be delayed upto a maximum of 3 weeks, without delaying the completion of the whole project.

Problem 2

Find out the completion time and the critical activities for the following project:

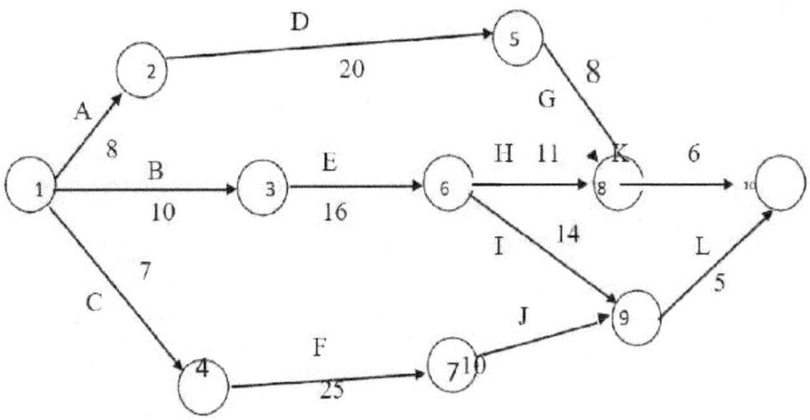

Solution

In all, we identify 4 paths, beginning with the start node of 1 and terminating at the end node of 10. They are as follows:

Path I

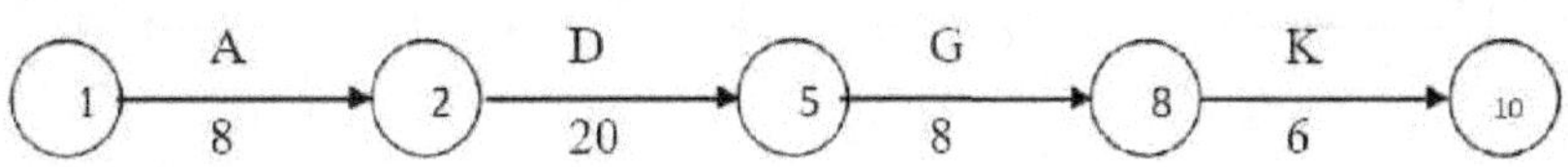

Time for the path = 8 + 20 + 8 + 6 = 42 units of time.

Path II

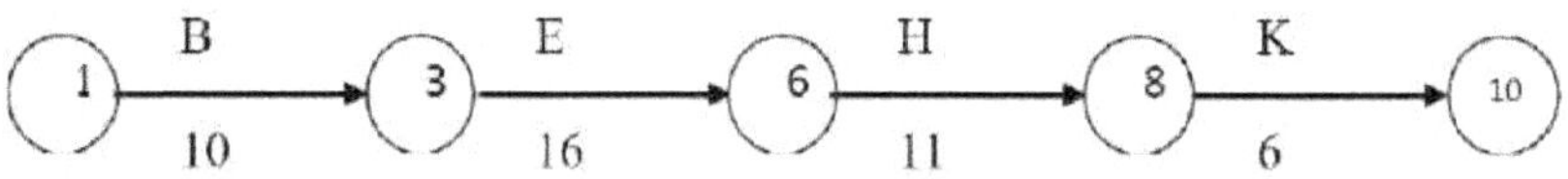

Time for the path = 10 + 16 + 11 + 6 = 43 units of time.

Path III

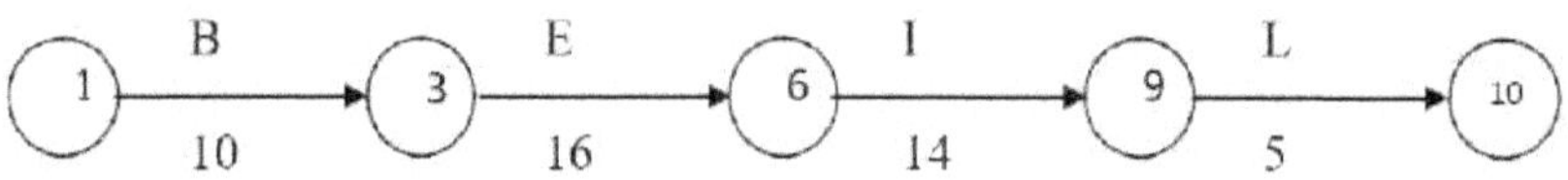

Time for the path = 10 + 16 + 14 + 5 = 45 units of time.

Path IV

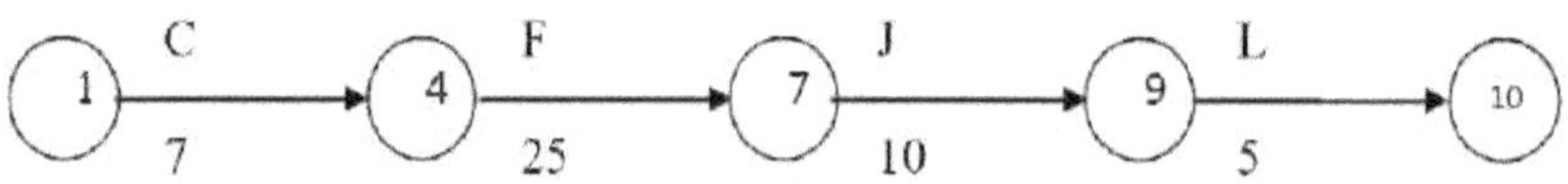

Time for the path = 7 + 25 + 10 + 5 = 47 units of time.

Compare the times for the four paths. Maximum of {42, 43, 45, 47} = 47. We see that the following path has the maximum time and so it is the critical path:

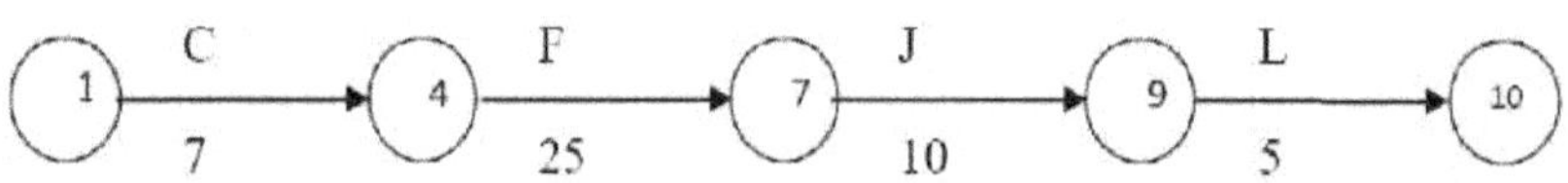

The critical activities are C, F, J and L. The non-critical activities are A, B, D, E, G, H, I and K. The project completion time is 47 units of time.

Problem 3

Draw the network diagram and determine the critical path for the following project:

Activity	Time estimate (Weeks)
1- 2	5
1- 3	6
1- 4	3
2 -5	5
3 -6	7
3 -7	10
4 -7	4
5 -8	2
6 -8	5
7 -9	6
8 -9	4

Solution

We have the following network diagram for the project:

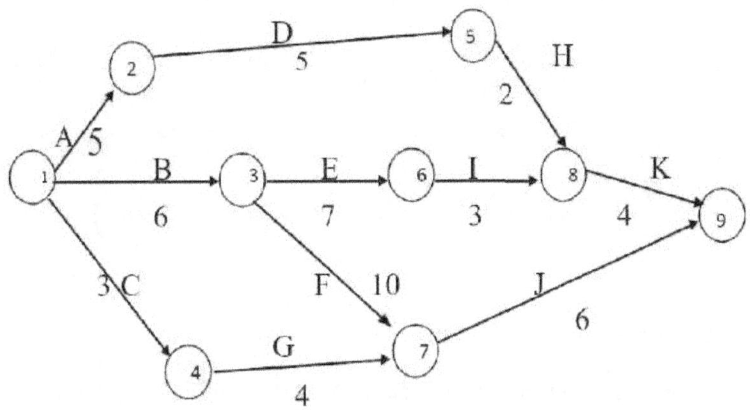

Solution

We assert that there are 4 paths, beginning with the start node of 1 and terminating at the end node of 9. They are as follows:

Path I

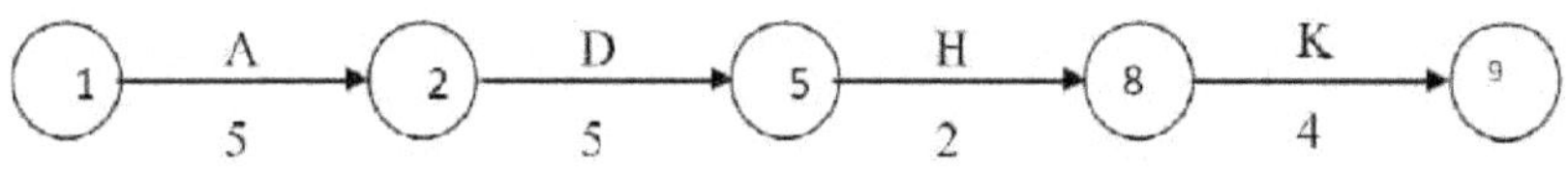

Time for the path = 5 + 5 + 2 + 4 = 16 weeks.

Path II

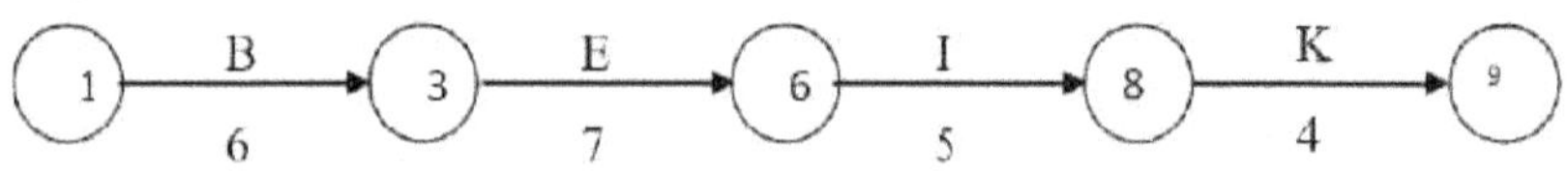

Time for the path = 6 + 7 + 5 + 4 = 22 weeks.

Path III

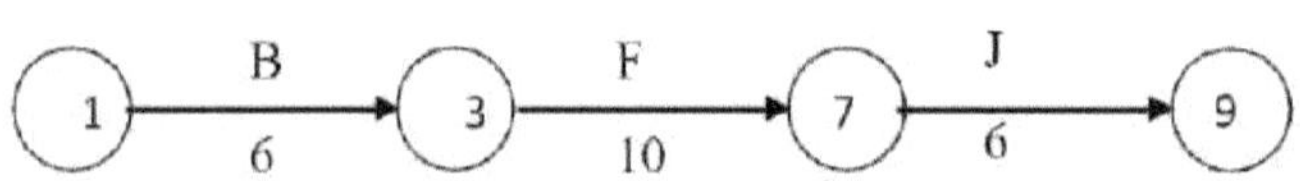

Time for the path = 6 + 10 + 6 = 16 weeks.

Path IV

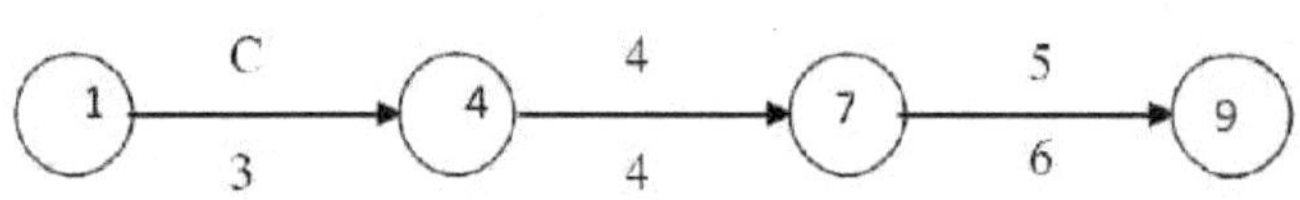

Time for the path = 3 + 4 + 6 = 13 weeks.

Compare the times for the four paths. Maximum of {16, 22, 16, 13} = 22. We see that the following path has the maximum time and so it is the critical path:

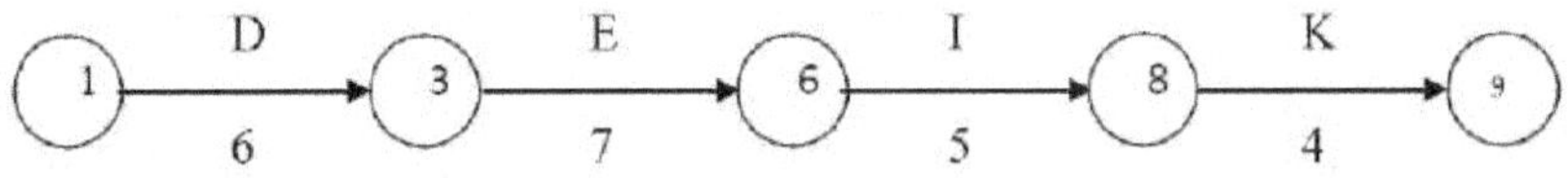

The critical activities are B, E, I and K. The non-critical activities are A, C, D, F, G, H and J. The project completion time is 22 weeks.

Questions

1. Explain the terms: critical path, critical activities.
2. The following are the time estimates and the precedence relationships of the activities in a project network:

Activity	IMMEDIATE Predecessor Activity	time estimate (weeks)
A	-	4
B	-	7
C	-	3
D	A	6
E	B	4
F	B	7
G	C	6
H	E	10
I	D	3
J	F, G	4
K	H, I	2

Draw the project network diagram. Determine the critical path and the project completion time.

****

Pert

Lesson Outline

- The Concept Of Pert
- Estimates Of The Time Of An Activity
- Determination Of Critical Path
- Probability Estimates
- Normal Probability Distribution Table

Learning Objectives

After reading this lesson you should be able to

- Understand the importance of PERT
- Locate the critical path
- Determine the project completion time
- Find out the probability of completion of a project before a stipulated time
- Use the normal probability distribution table

Introduction

Programme Evaluation and Review Technique (PERT) is a tool that would help a project manager in project planning and control. It would enable him in continuously monitoring a project and taking corrective measures wherever necessary. This technique involves statistical methods.

Assumptions for Pert

Note that in CPM, the assumption is that precise time estimate is available for each activity in a project. However, one finds most of the times that this is not practically possible.

In PERT, we assume that it is not possible to have precise time estimate for each activity and instead, probabilistic estimates of time alone are possible. A multiple time estimate approach is followed here. In probabilistic time estimate, the following 3 types of estimate are possible:

Pessimistic time estimate (t_p)

Optimistic time estimate (t_o)

Most likely time estimate (t_m)

The optimistic estimate of time is based on the assumption that an activity will not involve any difficulty during execution and it can be completed within a short period. On the other hand, a pessimistic estimate is made on the assumption that there would be unexpected problems during the execution of an activity and hence it would consume more time. The most likely time estimate is made in between the optimistic and the pessimistic estimates of time. Thus the three estimates of time have the relationship

$$t_o \leq t_m \leq t_p$$

Practically speaking, neither the pessimistic nor the optimistic estimate may hold in reality and it is the most likely time estimate that is expected to prevail in almost all cases. Therefore, it is preferable to give more weight to the most likely time estimate.

We give a weight of 4 to most likely time estimate and a weight of 1 each to the pessimistic and optimistic time estimates. We arrive at a time estimate (t_e) as the weighted average of these estimates as follows:

$$t_e = \frac{t_o + 4t_m + t_p}{6}$$

Since we have taken 6 units (1 for t_p , 4 for t_m and 1 for t_o), we divide the sum by 6. With this time estimate, we can determine the project completion time as applicable for CPM.

Since PERT involves the average of three estimates of time for each activity, this method is very practical and the results from PERT will be have a reasonable amount of reliability.

Measure Of Certainty

The 3 estimates of time are such that

$$t_o \leq t_m \leq t_p$$

Therefore the range for the time estimate is $t_p - t_o$.

The time taken by an activity in a project network follows a distribution with a standard deviation of one sixth of the range, approximately.

i.e., The standard deviation = $\sigma = \frac{t_p - t_o}{6}$

and the variance = $\sigma^2 = \left(\frac{t_p - t_o}{6}\right)^2$

The certainty of the time estimate of an activity can be analysed with the help of the variance. The greater the variance, the more uncertainty in the time estimate of an activity.

Problem 1

Two experts A and B examined an activity and arrived at the following time estimates.

Expert	Time Estimate		
	t_o	t_m	t_p
A	4	6	8
B	4	7	10

Determine which expert is more certain about his estimates of time:

Solution

Variance (σ^2) in time estimates = $\left(\frac{t_p - t_o}{6}\right)^2$

In the case of expert A, the variance = $\left(\frac{8-4}{6}\right)^2 = \frac{4}{9}$

As regards expert B, the variance = $\left(\frac{10-4}{6}\right)^2 = 1$

So, the variance is less in the case of A. Hence, it is concluded that the expert A is more certain about his estimates of time.

Determination of Project Completion Time in PERT

Problem 2

Find out the time required to complete the following project and the critical activities:

Activity	Predecessor Activity	Optimistic time estimate (to days)	Most likely time estimate (tm days)	Pessimistic time estimate (tp days)
A	-	2	4	6
B	A	3	6	9
C	A	8	10	12
D	B	9	12	15
E	C	8	9	10
F	D, E	16	21	26
G	D, E	19	22	25
H	F	2	5	8
I	G	1	3	5

Solution

From the three time estimates t_p, t_m and t_o, calculate t_e for each activity. We obtain the following table:

Activity	Optimistic time estimate (to)	4 x Most likely time estimate	Pessimistic time estimate (tp)	to+ 4tm + tp	Time estimate $t_e = \frac{t_o + 4t_m + t_p}{6}$
A	2	16	6	24	4
B	3	24	9	36	6
C	8	40	12	60	10
D	9	48	15	72	12
E	8	36	10	54	9
F	16	84	26	126	21
G	19	88	25	132	22
H	2	20	8	30	5
I	1	12	5	18	3

Using the single time estimates of the activities, we get the following network diagram for the project.

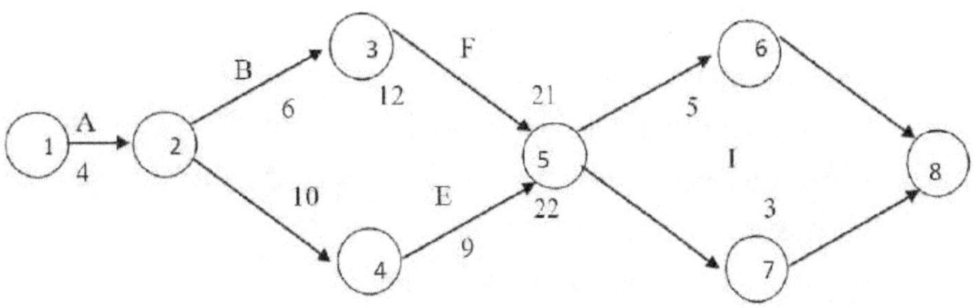

Consider the paths, beginning with the start node and stopping with the end node. There are four such paths for the given project. They are as follows:

Path I

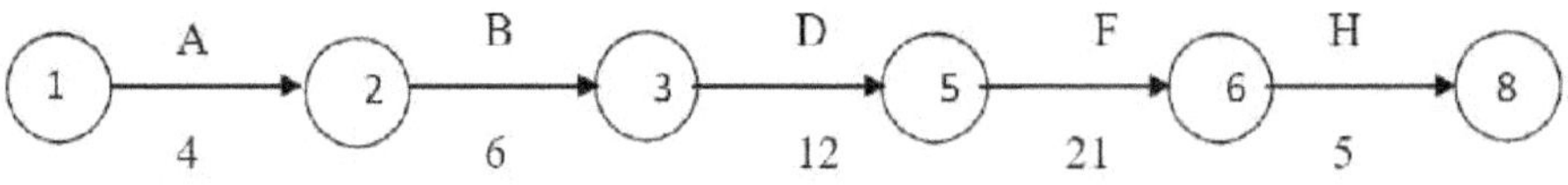

Time for the path: 4+6+12+21+5 = 48 days.

Path II

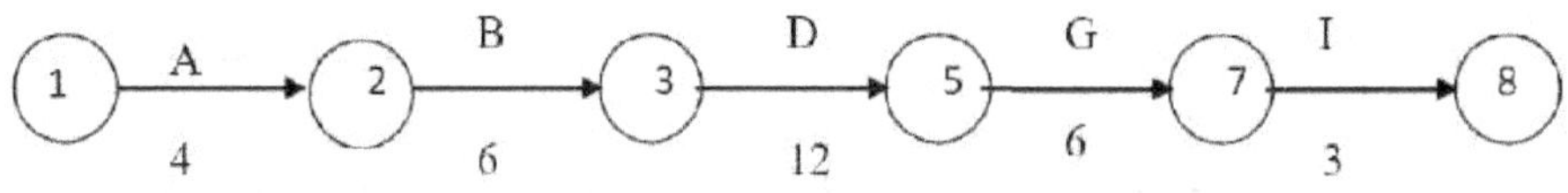

Time for the path: 4+6+12+ 6+3 = 31 days.

Path III

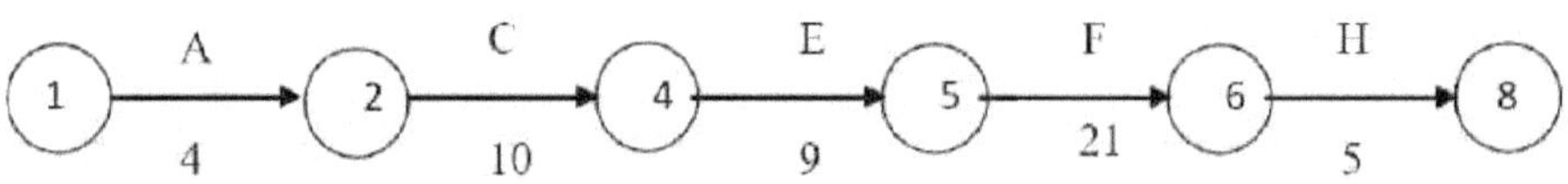

Time for the path: 4+10+9+ 21+5 = 49 days.

Path IV

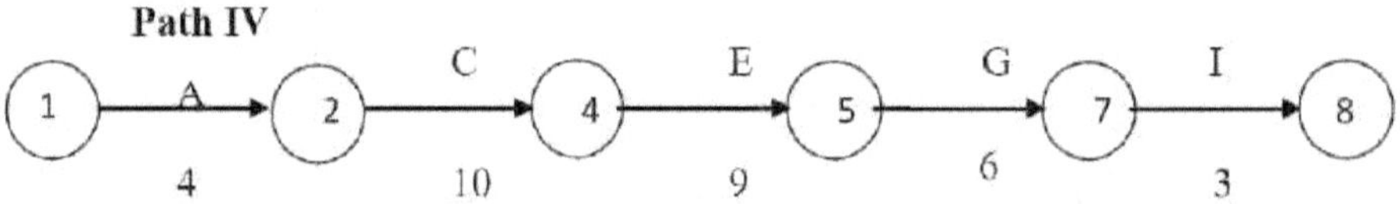

Time for the path: 4+10+9+ 6+3 = 32 days.

Compare the times for the four paths.

Maximum of {48, 31, 49, 32} = 49.

We see that Path III has the maximum time.

Therefore the critical path is Path III. i.e., 1→2→4→5→6→8.

The critical activities are A, C, E, F and H.

The non-critical activities are B, D, G and I.

Project time (Also called project length) = 49 days.

Problem 3

Find out the time, variance and standard deviation of the project with the following time estimates in weeks:

Activity	Optimistic time estimate (to)	Most likely time estimate (tm)	Pessimistic time estimate (tp)
1-2	3	6	9
1-6	2	5	8
2-3	6	12	18
2-4	4	5	6
3-5	8	11	14
4-5	3	7	11
6-7	3	9	15
5-8	2	4	6
7-8	8	16	18

Solution

From the three time estimates t_p , t_m and t_o , calculate t_e for each activity. We obtain the following table:

Activity	Optimistic time stimate (to)	4 x Most likely time estimate	Pessimistic time estimate (tp)	to+ 4tm + tp	Time estimate $t_e = \frac{t_o + 4t_m + t_p}{6}$
1-2	3	24	9	36	6
1-6	2	20	8	30	5
2-3	6	48	18	72	12
2-4	4	20	6	30	5
3-5	8	44	14	66	11

4-5	3	28	11	42	7
6-7	3	36	15	54	9
5-8	2	16	6	24	4
7-8	8	64	18	90	15

With the single time estimates of the activities, we get the following network diagram for the project.

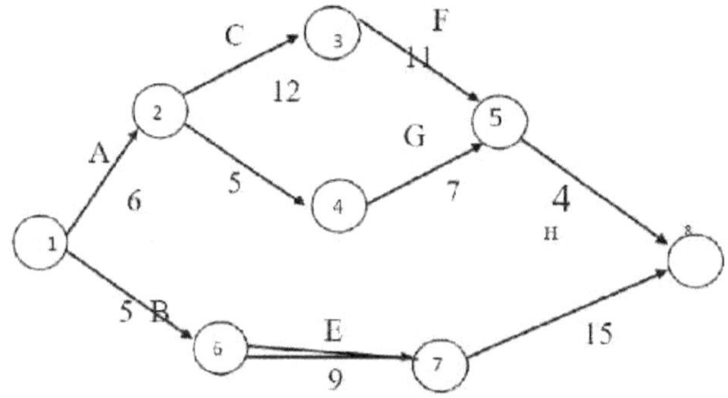

Consider the paths, beginning with the start node and stopping with the end node. There are three such paths for the given project. They are as follows:

Path I

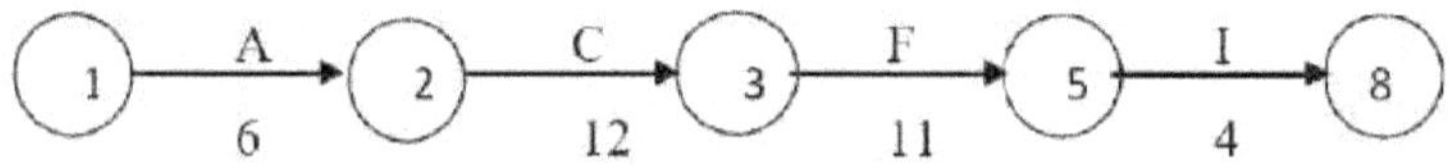

Time for the path: 6+12+11+4 = 33 weeks.

Path II

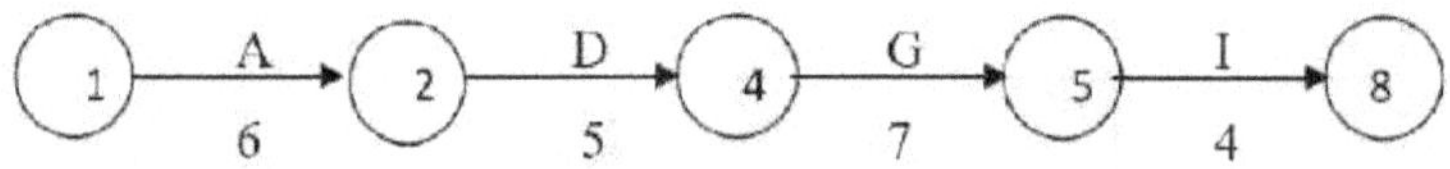

Time for the path: 6+5+7+ 4= 22 weeks.

Path III

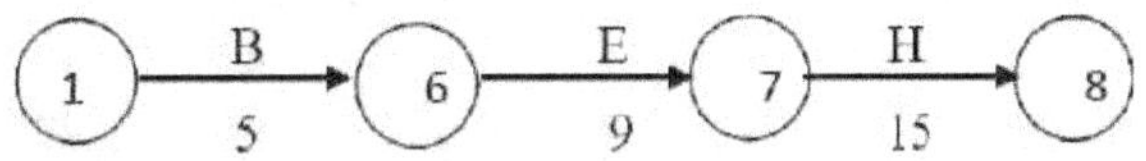

Time for the path: 5+9+15 = 29 weeks.

Compare the times for the three paths.

Maximum of {33, 22, 29} = 33.

It is noticed that Path I has the maximum time.

Therefore the critical path is Path I. i.e., 1→2→3→5→ 8

The critical activities are A, C, F and I.

The non-critical activities are B, D, G and H.

Project time = 33 weeks.

Calculation of Standard Deviation and Variance for the Critical Activities:

Critical Activity	Optimistic time estimate (to)	Most likely timeestimate (tm)	Pessimistic time estimate (tp)	Range (tp - to)	Standard deviation = $\sigma = \frac{t_p - t_o}{6}$	Variance $\sigma^2 = \left(\frac{t_p - t_o}{6}\right)^2$
A: 1 2	3	6	9	6	1	1
C: 2 3	6	12	18	12	2	4
F: 3 5	8	11	14	6	1	1
I: 5 8	2	4	6	4	2/3	4/9

Variance of project time (Also called Variance of project length) =

Sum of the variances for the critical activities = 1+4+1+ 4/9 = 58/9 Weeks.

Standard deviation of project time = √Variance = √58/9 = 2.54 weeks.

Problem 4

A project consists of seven activities with the following time estimates. Find the probability that the project will be completed in 30 weeks or less.

Activity	Predecessor Activity	Optimistic time estimate (to days)	Most likely time estimate (tm days)	Pessimistic time estimate (tp days)
A	-	2	5	8
B	A	2	3	4
C	A	6	8	10
D	A	2	4	6
E	B	2	6	10
F	C	6	7	8
G	D,E,F	6	8	10

Solution

From the three time estimates , and , calculate for each activity. The results are furnished in the following table:

Activity	Optimistic time estimate (to)	4 x Most likely time estimate	Pessimistic time estimate (tp)	to+ 4tm + tp	Time estimate $t_e = \frac{t_o + 4t_m + t_p}{6}$
A	2	20	8	30	5
B	2	12	4	18	3
C	6	32	10	48	8
D	2	16	6	24	4
E	2	24	10	36	6
F	6	28	8	42	7
G	6	32	10	48	8

With the single time estimates of the activities, the following network diagram is constructed for the project.

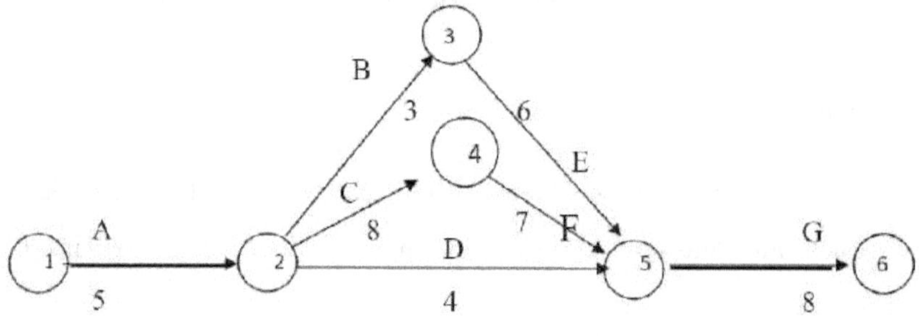

Consider the paths, beginning with the start node and stopping with the end node. There are three such paths for the given project. They are as follows:

Path I

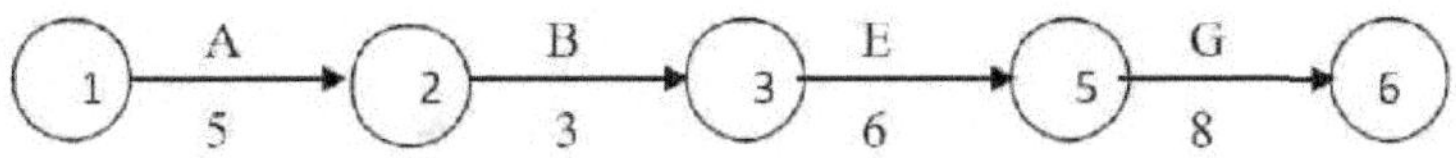

Time for the path: 5+3+6+8 = 22 weeks.

Path II

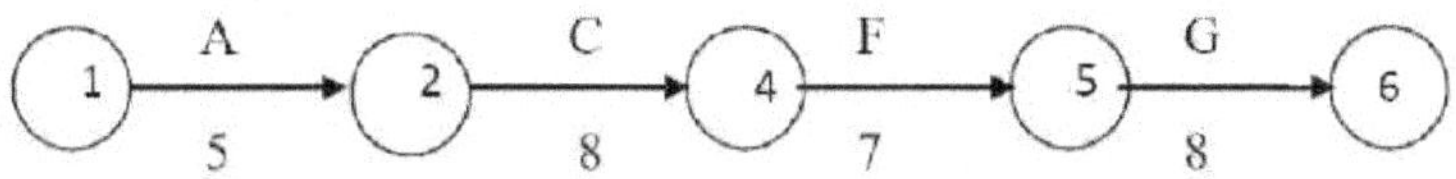

Time for the path: 5+8+7+ 8 = 28 weeks.

Path III

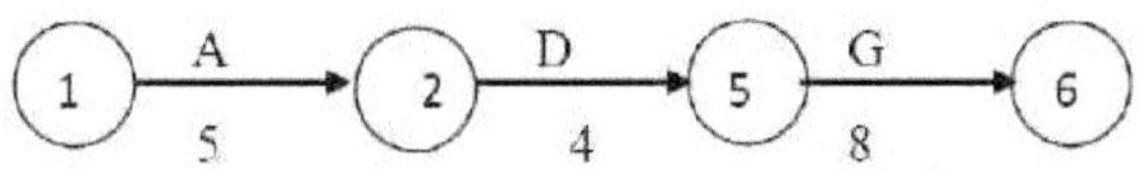

Time for the path: 5+4+8 = 17 weeks.

Compare the times for the three paths.

Maximum of {22, 28, 17} = 28.

It is noticed that Path II has the maximum time.

Therefore the critical path is Path II. i.e., 1→2→ 4→5→ 6.

The critical activities are A, C, F and G.

The non-critical activities are B, D and E.

Project time = 28 weeks.

Calçulation of Standard Deviation and Variance for the

Critical Activities:

Critical Activity	Optimistic time estimate (to)	Most likely time estimate (tm)	Pessimistic time estimate (tp)	Range (tp -to)	Standard deviation = $\sigma = \frac{t_p - t_o}{6}$	Variance $\sigma^2 = \left(\frac{t_p - t_o}{6}\right)^2$
A: 1 2	2	5	8	6	1	1
C: 2 4	6	8	10	4	$\frac{2}{3}$	$\frac{4}{9}$
F: 4 5	6	7	8	2	$\frac{1}{3}$	$\frac{1}{9}$
G: 5 6	6	8	10	4	$\frac{2}{3}$	$\frac{4}{9}$

Standard deviation of the critical path = √2 = 1.414

The standard normal variate is given by the formula

$$Z = \frac{\textit{Given value of t} - \textit{Expected value of t in the critical path}}{\textit{SD for the critical path}}$$

$$\text{So we get } Z = \frac{30 - 28}{1.414} = 1.414$$

We refer to the Normal Probability Distribution Table.

Corresponding to Z = 1.414, we obtain the value of 0.4207

We get 0.5 + 0.4207 = 0. 9207

Therefore the required probability is 0.92

i.e., There is 92% chance that the project will be completed before 30 weeks. In other words, the chance that it will be delayed beyond 30 weeks is 8%

Questions

1. Explain how time of an activity is estimated in PERT.
2. Explain the measure of certainty in PERT.
3. The estimates of time in weeks of the activities of a project are as follows:

Activity	Predecessor Activity	Optimistic estimate of time	Most likely estimate of time	Pessimistic estimate of time
A	-	2	4	6
B	A	8	11	20
C	A	10	15	20
D	B	12	18	24
E	C	8	13	24
F	C	4	7	16
G	D,F	14	18	28
H	E	10	12	14
I	G,H	7	10	19

Determine the critical activities and the project completion time.

4. Draw the network diagram for the following project. Determine the time, variance and standard deviation of the project.:

Activity	Predecessor Activity	Optimistic estimate of time	Most likely estimate of time	Pessimistic estimate of time
A	-	12	14	22
B	-	16	17	24
C	A	14	15	16
D	A	13	18	23
E	B	16	18	20
F	D,E	13	14	21
G	C,F	6	8	10

5. Consider the following project with the estimates of time in weeks:

Activity	Predecessor Activity	Optimistic estimate of time	Most likely estimate of time	Pessimistic estimate of time
A	-	2	4	6
B	-	3	5	7
C	A	5	6	13
D	A	4	8	12
E	B,C	5	6	13
F	D,E	6	8	14

Find the probability that the project will be completed in 27 weeks.

NORMAL DISTRIBUTION TABLE

Area Under Standard Normal Distribution

	0.00	0.01	0.02	0.03	0.04	0.05	0.06	0.07	0.08	0.09
0.0	0.0000	0.0040	0.0080	0.0120	0.0160	0.0199	0.0239	0.0279	0.0319	0.0359
0.1	0.0398	0.0438	0.0478	0.0517	0.0557	0.0596	0.0636	0.0675	0.0714	0.0753
0.2	0.0793	0.0832	0.0871	0.0910	0.0948	0.0987	0.1026	0.1064	0.1103	0.1141
0.3	0.1179	0.1217	0.1255	0.1293	0.1331	0.1368	0.1406	0.1443	0.1480	0.1517
0.4	0.1554	0.1591	0.1628	0.1664	0.1700	0.1736	0.1772	0.1808	0.1844	0.1879
0.5	0.1915	0.1950	0.1985	0.2019	0.2054	0.2088	0.2123	0.2157	0.2190	0.2224
0.6	0.2257	0.2291	0.2324	0.2357	0.2389	0.2422	0.2454	0.2486	0.2517	0.2549
0.7	0.2580	0.2611	0.2642	0.2673	0.2704	0.2734	0.2764	0.2794	0.2823	0.2852
0.8	0.2881	0.2910	0.2939	0.2967	0.2995	0.3023	0.3051	0.3078	0.3106	0.3133
0.9	0.3159	0.3186	0.3212	0.3238	0.3264	0.3289	0.3315	0.3340	0.3365	0.3389
1.0	0.3413	0.3438	0.3461	0.3485	0.3508	0.3531	0.3554	0.3577	0.3599	0.3621
1.1	0.3643	0.3665	0.3686	0.3708	0.3729	0.3749	0.3770	0.3790	0.3810	0.3830
1.2	0.3849	0.3869	0.3888	0.3907	0.3925	0.3944	0.3962	0.3980	0.3997	0.4015
1.3	0.4032	0.4049	0.4066	0.4082	0.4099	0.4115	0.4131	0.4147	0.4162	0.4177
1.4	0.4192	0.4207	0.4222	0.4236	0.4251	0.4265	0.4279	0.4292	0.4306	0.4319
1.5	0.4332	0.4345	0.4357	0.4370	0.4382	0.4394	0.4406	0.4418	0.4429	0.4441
1.6	0.4452	0.4463	0.4474	0.4484	0.4495	0.4505	0.4515	0.4525	0.4535	0.4545
1.7	0.4554	0.4564	0.4573	0.4582	0.4591	0.4599	0.4608	0.4616	0.4625	0.4633
1.8	0.4641	0.4649	0.4656	0.4664	0.4671	0.4678	0.4686	0.4693	0.4699	0.4706
1.9	0.4713	0.4719	0.4726	0.4732	0.4738	0.4744	0.4750	0.4756	0.4761	0.4767
2.0	0.4772	0.4778	0.4783	0.4788	0.4793	0.4798	0.4803	0.4808	0.4812	0.4817

2.1	0.4821	0.4826	0.4830	0.4834	0.4838	0.4842	0.4846	0.4850	0.4854	0.4857
2.2	0.4861	0.4864	0.4868	0.4871	0.4875	0.4878	0.4881	0.4884	0.4887	0.4890
2.3	0.4893	0.4896	0.4898	0.4901	0.4904	0.4906	0.4909	0.4911	0.4913	0.4916
2.4	0.4918	0.4920	0.4922	0.4925	0.4927	0.4929	0.4931	0.4932	0.4934	0.4936
2.5	0.4938	0.4940	0.4941	0.4943	0.4945	0.4946	0.4948	0.4949	0.4951	0.4952
2.6	0.4953	0.4955	0.4956	0.4957	0.4959	0.4960	0.4961	0.4962	0.4963	0.4964
2.7	0.4965	0.4966	0.4967	0.4968	0.4969	0.4970	0.4971	0.4972	0.4973	0.4974
2.8	0.4974	0.4975	0.4976	0.4977	0.4977	0.4978	0.4979	0.4979	0.4980	0.4981
2.9	0.4981	0.4982	0.4982	0.4983	0.4984	0.4984	0.4985	0.4985	0.4986	0.4986
3.0	0.4987	0.4987	0.4987	0.4988	0.4988	0.4989	0.4989	0.4989	0.4990	0.4990

****

Earliest And Latest Times

Lesson Outline

- The Concepts Of Earliest And Latest Times
- The Concept Of Slack
- Numerical Problems

Learning Objectives

- After reading this lesson you should be able to
- Understand the concepts of earliest and latest times
- Understand the concept of slack
- Calculate the earliest and latest times
- Find out the slacks
- Identify the critical activities
- Carry out numerical problems

Introduction

A project manager has the responsibility to see that a project is completed by the stipulated date, without delay. Attention is focused on this aspect in what follows.

Key concepts

Certain key concepts are introduced below.

Earliest Times Of An Activity

We can consider (i) Earliest Start Time of an activity and (ii) Earliest Finish Time of an activity.

Earliest Start Time of an activity is the earliest possible time of starting that activity on the condition that all the other activities preceding to it were began at the earliest possible times.

Earliest Finish Time of an activity is the earliest possible time of completing that activity. It is given by the formula.

The Earliest Finish Time of an activity = The Earliest Start Time of the activity + The estimated duration to carry out that activity.

LATEST TIMES OF AN ACTIVITY

We can consider (i) Latest Finish Time of an activity and (ii) Latest Start Time of an activity.

Latest Finish Time of an activity is the latest possible time of completing that activity on the condition that all the other activities succeeding it are carried out as per the plan of the management and without delaying the project beyond the stipulated time.

Latest Start Time of an activity is the latest possible time of beginning that activity. It is given by the formula

Latest Start Time of an activity = The Latest Finish Time of the activity - The estimated duration to carry out that activity.

TOTAL FLOAT OF AN ACTIVITY

Float seeks to measure how much delay is acceptable. It sets up a control limit for delay.

The total float of an activity is the time by which that activity can be delayed without delaying the whole project. It is given by the formula

Total Float of an Activity = Latest Finish Time of the activity - Earliest Finish Time of that activity.

It is also given by the formula

Total Float of an Activity = Latest Start Time of the activity - Earliest

Start Time of that activity.

Since a delay in a critical activity will delay the execution of the whole project, the total float of a critical activity must be zero.

EXPECTED TIMES OF AN EVENT

An event occurs at a point of time. We can consider (i) Earliest Expected Time of Occurrence of an event and (ii) Latest Allowable Time of Occurrence an event.

The Earliest Expected Time of Occurrence of an event is the earliest possible time of expecting that event to happen on the condition that all the preceding activities have been completed.

The Latest Allowable Time of Occurrence of an event is the latest possible time of expecting that event to happen without delaying the project beyond the stipulated time.

Procudure To Find The Earliest Expected Time Of An Event

Step 1

Take the Earliest Expected Time of Occurrence of the Start Event as zero.

Step 2

For an event other than the Start Event, find out all paths in the network which connect the Start node with the node representing the event under consideration.

Step 3

In the "Forward Pass" (i.e., movement in the network from left to right), find out the sum of the time durations of the activities in each path identified in Step 2.

Step 4.

The path with the longest time in Step 3 gives the Earliest Expected Time of Occurrence of the event

Working Rule for finding the earliest expected time of an event:

For an event under consideration, locate all the predecessor events and identify their earliest expected times. With the earliest expected time of each event, add the time duration of the activity connecting that event to the event under consideration. The maximum among all these values gives the Earliest Expected Time of Occurrence of the event.

Procudure To Find The Latest Allowable Time Of An Event

We consider the "Backward Pass" (i.e., movement in the network from right to left).

The latest allowable time of occurrence of the End Node must be the time of completion of the project. Therefore it shall be equal to the time of the critical path of the project.

Step 1

Identify the latest allowable time of occurrence of the End Node.

Step 2

For an event other than the End Event, find out all paths in the network which connect the End node with the node representing the event under consideration.

Step 3

In the "Backward Pass" (i.e., movement in the network from right to left), subtract the time durations of the activities along each such path.

Step 4

The Latest Allowable Time of Occurrence of the event is determined by the path with the longest time in Step 3. In other words, the smallest value of time obtained in Step 3 gives the Latest Allowable Time of Occurrence of the event.

Working Rule for finding the latest allowable time of an event:

For an event under consideration, locate all the successor events and identify their latest allowable times. From the latest allowable time of each successor event, subtract the time duration of the activity that begins with the event under consideration. The minimum among all these values gives the Latest Allowable Time of Occurrence of the event.

Slack Of An Event

The allowable time gap for the occurrence of an event is known as the slack of that event. It is given by the formula

Slack of an event = Latest Allowable Time of Occurrence of the event - Earliest Expected Time of Occurrence of that event.

Slack Of An Activity

The slack of an activity is the float of the activity.

Problem 1

The following details are available regarding a project:

Activity	Predecessor Activity	Duration (Weeks)
A	-	12
B	A	7
C	A	11
D	A	8
E	A	6
F	B	10
G	C	9
H	D, F	14
I	E, G	13
J	H, I	16

Determine the earliest and latest times, the total float for each activity, the critical activities and the project completion time.

Solution

With the given data, we construct the following network diagram for the project.

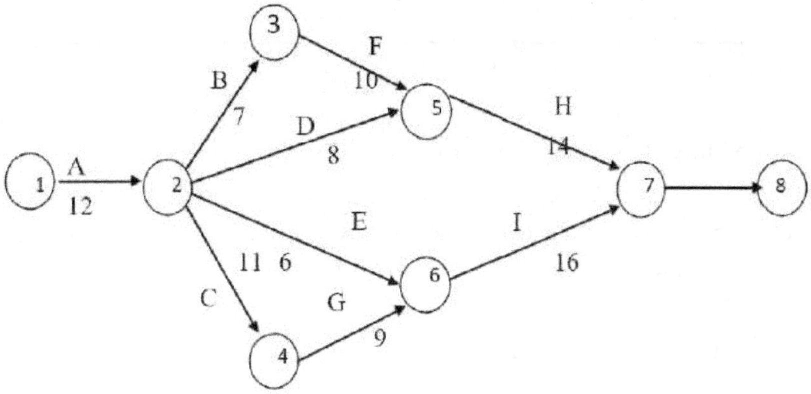

Consider the paths, beginning with the start node and stopping with the end node. There are four such paths for the given project. They are as follows:

Path I

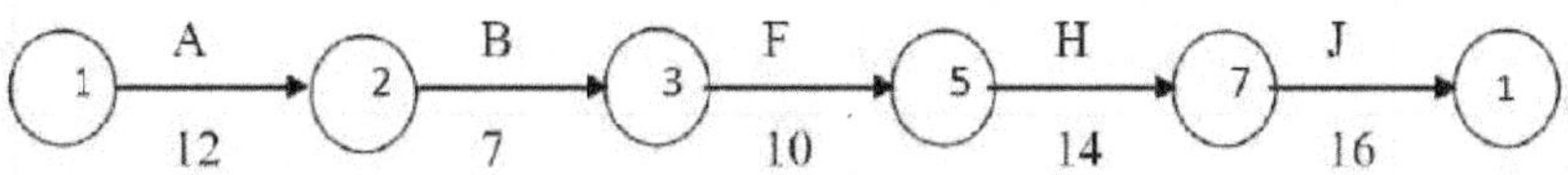

Time of the path = 12 + 7 + 10 + 14 + 16 = 59 weeks.

Path II

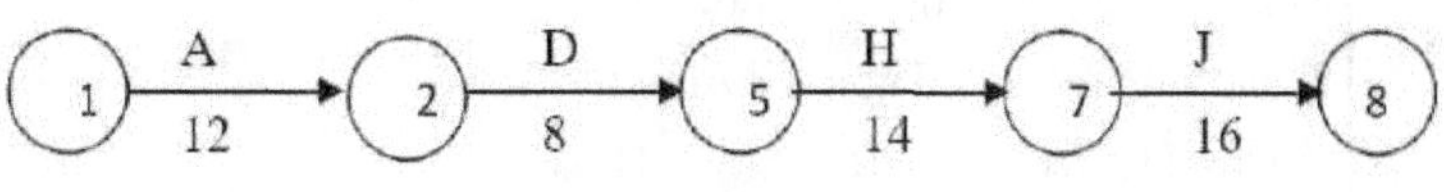

Time of the path = 12 + 8 + 14 + 16 = 50 weeks.

Path III

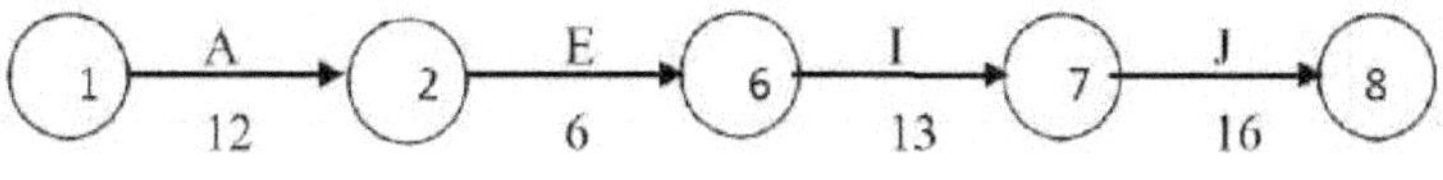

Time of the path = 12 + 6 + 13 + 16 = 47 weeks.

Path IV

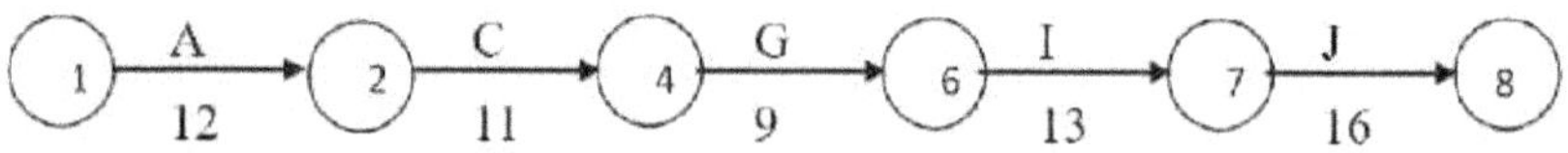

Time of the path = 12 + 11 + 9 + 13 + 16 = 61 weeks.

Compare the times for the four paths. Maximum of {51, 50, 47, 61} = 61.
We see that the maximum time of a path is 61 weeks.

Forward pass

Calculation of Earliest Expected Time of Occurrence of Events

Node	Earliest Time of Occurrence of Node
1	0
2	Time for Node 1 + Time for Activity A = 0 + 12 = 12
3	Time for Node 2 + Time for Activity B = 12 + 7 = 19
4	Time for Node 2 + Time for Activity C = 12 + 11 = 23
5	Max {Time for Node 2 + Time for Activity D, Time for Node 3 + Time for Activity F} = Max {12 + 8, 19 + 10} = Max {20, 29} = 29
6	Max {Time for Node 2 + Time for Activity E, Time for Node 4 + Time for Activity G} = Max {12 + 6, 23 + 9} = Max {18, 32} = 32
7	Max {Time for Node 5 + Time for Activity H, Time for Node 6 + Time for Activity I} = Max {29 + 14, 32 + 13} = Max {43, 45} = 45
8	Time for Node 7 + Time for Activity J = 45 + 16 = 61

Using the above values, we obtain the Earliest Start Times of the activities as follows:

Activity	Earliest Start Time (Weeks)
A	0
B	12
C	12
D	12
E	12
F	19
G	23
H	29
I	32
J	45

Backward pass

Calculation of Latest Allowable Time of Occurrence of Events

Node	**Latest Allowable Time of Occurrence of Node**
8	Maximum time of a path in the network = 61
7	Time for Node 8 - Time for Activity J = 61 -16 = 45
6	Time for Node 7 - Time for Activity I = 45 -13 = 32
5	Time for Node 7 - Time for Activity H = 45 -14 = 31
4	Time for Node 6 - Time for Activity G = 32 - 9 = 23
3	Time for Node 5 - Time for Activity F = 31- 10 = 21

2	Min {Time for Node 3 - Time for Activity B, Time for Node 4 - Time for Activity C, Time for Node 5 - Time for Activity D, Time for Node 6 - Time for Activity E} = Min {21 - 7, 23-11,31-8, 32-6} = Min {14, 12, 23, 26} = 12
1	Time for Node 2 - Time for Activity A = 12- 12 = 0

Using the above values, we obtain the Latest Finish Times of the activities as follows:

Activity	Latest Finish Time (Weeks)
J	61
I	45
H	45
G	32
F	31
E	32
D	31
C	23
B	21
A	12

Calculation of Total Float for each activity:

Activity	Duration (Weeks)	Earliest Start Time	Earliest Finish Time	Latest Start Time	Latest Finish Time	Total Float = Latest Finish Time - Earliest Finish Time
A	12	0	12	0	12	0
B	7	12	19	14	21	2
C	11	12	23	12	23	0
D	8	12	20	23	31	11
E	6	12	18	26	32	14
F	10	19	29	21	31	2
G	9	23	32	23	32	0
H	14	29	43	31	45	2
I	13	32	45	32	45	0
J	16	45	61	45	61	0

The activities with total float = 0 are A, C, G, I and J. They are the critical activities.

Project completion time = 61 weeks.

Problem 2

The following are the details of the activities in a project:

Activity	Predecessor Activity	Duration (Weeks)
A	-	15
B	A	17
C	A	21

D	B	19
E	B	22
F	C, D	18
G	E, F	15

Calculate the earliest and latest times, the total float for each activity and the project completion time.

Solution

The following network diagram is obtained for the given project.

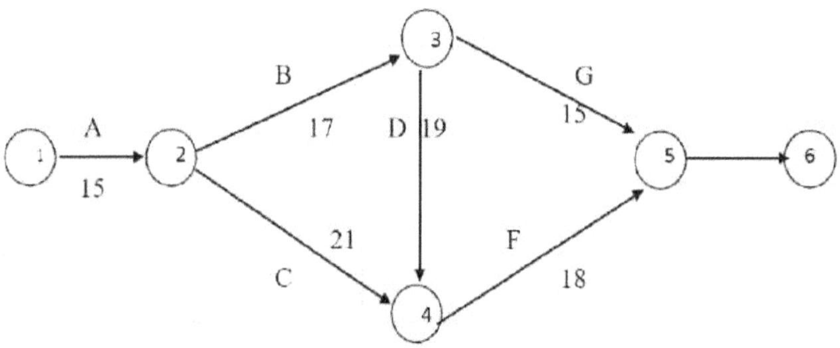

Consider the paths, beginning with the start node and stopping with the end node. There are three such paths for the given project. They are as follows:

Path I

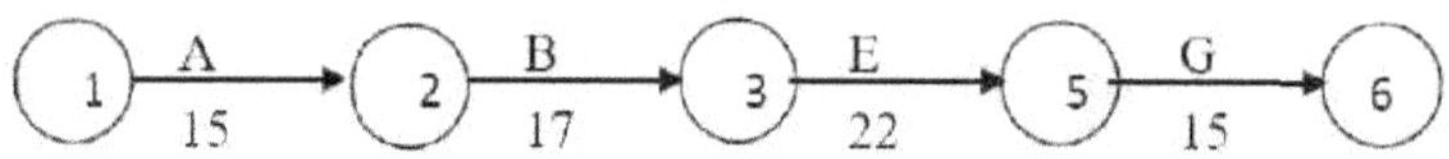

Time of the path = 15 + 17 + 22 + 15 = 69 weeks.

Path II

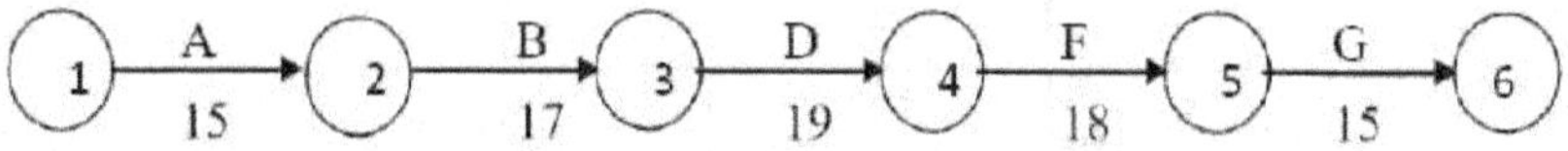

Time of the path = 15 + 17 + 19 + 18 + 15 = 84 weeks.

Path III

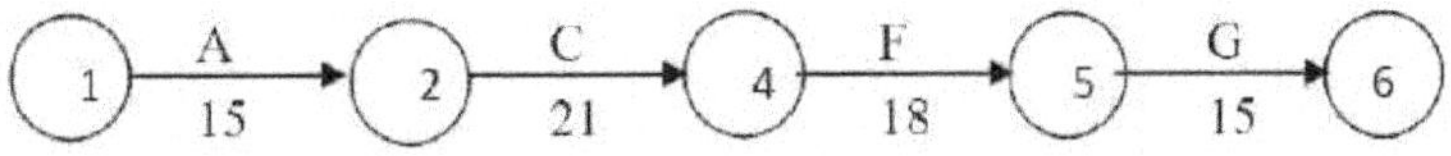

Time of the path = 15 + 21 + 18 + 15 = 69 weeks.

Compare the times for the three paths. Maximum of {69, 84, 69} = 84. We see that the maximum time of a path is 84 weeks.

Forward pass

Calculation of Earliest Time of Occurrence of Events

Node	Earliest Time of Occurrence of Node
1	0
2	Time for Node 1 + Time for Activity A = 0 + 15 = 15
3	Time for Node 2 + Time for Activity B = 15 + 17 = 32
4	Max {Time for Node 2 + Time for Activity C, Time for Node 3 + Time for Activity D} = Max {15 + 21, 32 + 19} = Max {36, 51} = 51
5	Max {Time for Node 3 + Time for Activity E, Time for Node 4 + Time for Activity F} = Max {32 + 22, 51 + 18} = Max {54, 69} = 69
6	Time for Node 5 + Time for Activity G = 69 + 15 = 84

Calculation of Earliest Time for Activities

Activity	Earliest Start Time (Weeks)
A	0
B	15
C	15
D	32
E	32
F	51
G	69

Backward pass

Calculation of the Latest Allowable Time of Occurrence of Events

Node	Latest Allowable Time of Occurrence of Node
6	Maximum time of a path in the network = 84
5	Time for Node 6 - Time for Activity G = 84 -15 = 69
4	Time for Node 5 - Time for Activity F = 69 - 18 = 51
3	Min {Time for Node 4 - Time for Activity D, Time for Node 5 - Time for Activity E} = Min {51 - 19, 69 - 22} = Min {32, 47} = 32
2	Min {Time for Node 3 - Time for Activity B, Time for Node 4 - Time for Activity C} = Min {32 - 17, 51 - 21} = Min {15, 30} = 15
1	Time for Node 2 - Time for Activity A = 15 - 15 = 0

Calculation of the Latest Finish Times of the activities

Activity	Latest Finish Time (Weeks)
G	84
F	69
E	69
D	51
C	51
B	32
A	15

Calculation of Total Float for each activity:

Activity	Duration (Weeks)	Earliest Start Time	Earliest Finish Time	Latest Start Time	Latest Finish Time	Total Float = Latest Finish Time - Earliest Finish Time
A	15	0	15	0	15	0
B	17	15	32	15	32	0
C	21	15	36	30	51	15
D	19	32	51	32	51	0
E	22	32	54	47	69	15
F	18	51	69	51	69	0
G	15	69	84	69	84	0

The activities with total float = 0 are A, B, D, F and G. They are the critical activities.
Project completion time = 84 weeks.

Problem 3

Consider a project with the following details:

Name of Activity	Predecessor Activity	Duration (Weeks)
A	-	8
B	A	13
C	A	9
D	A	12
E	B	14
F	B	8
G	D	7
H	C,F,G	12
I	C,F,G	9
J	E, H	10
K	I, J	7

Determine the earliest and latest times, the total float for each activity, the critical activities, the slacks of the events and the project completion time.

Solution

The following network diagram is got for the given project:

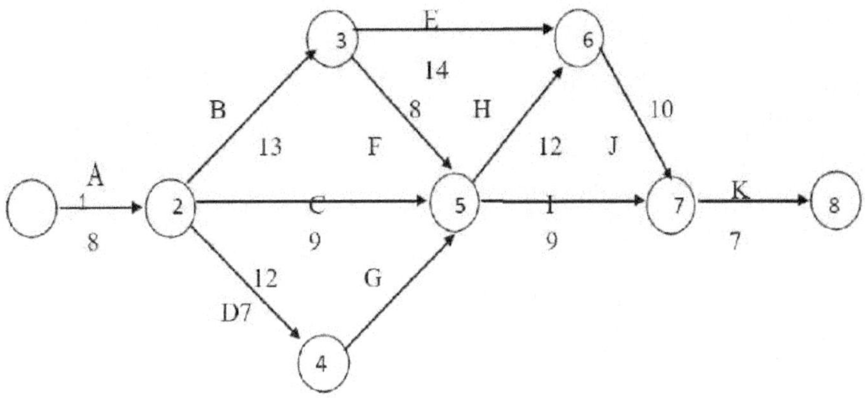

Path I

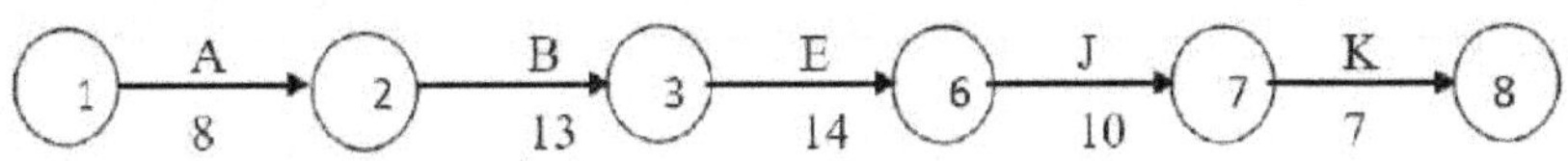

Time of the path = 8 + 13 + 14 + 10 + 7 = 52 weeks.

Path II

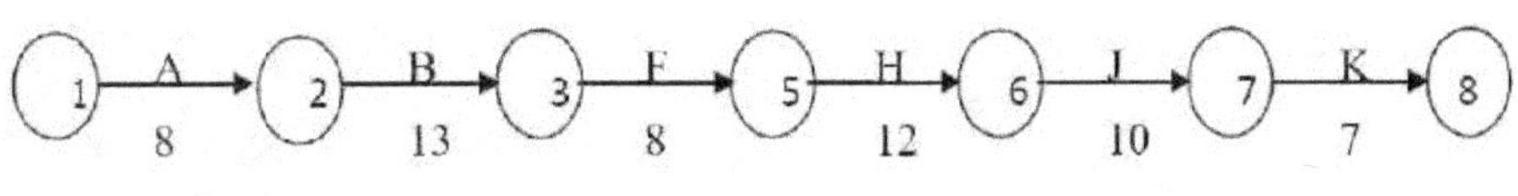

Time of the path = 8 + 13 + 8 + 12 + 10 + 7 = 58 weeks.

Path III

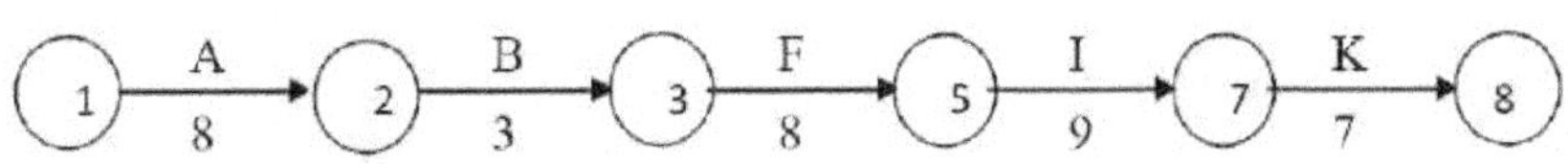

Time of the path = 8 + 13 + 8 + 9 + 7 = 45 weeks.

Path IV

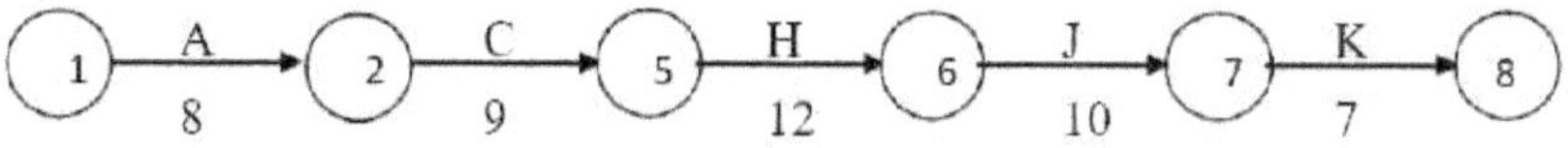

Time of the path = 8 + 9 + 12 + 10 + 7 = 46 weeks.

Path V

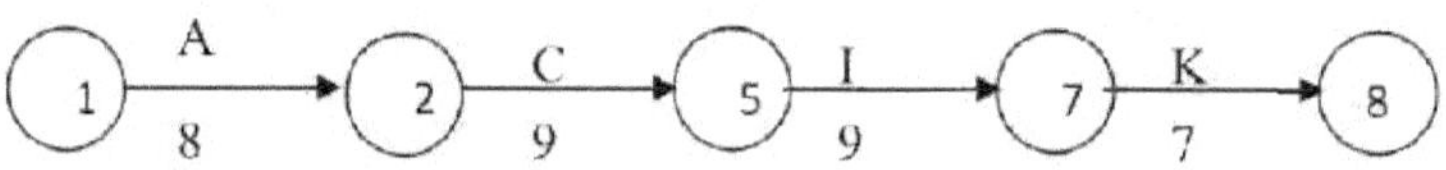

Time of the path = 8 + 9 + 9 + 7 = 33 weeks.

Path VI

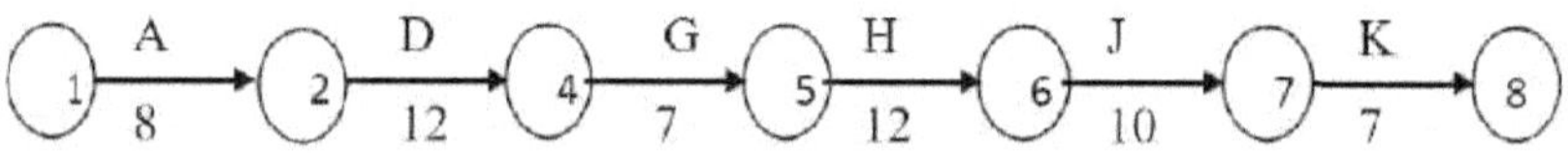

Time of the path = 8 + 12 + 7 + 12 + 10 + 7 = 56 weeks.

Path VII

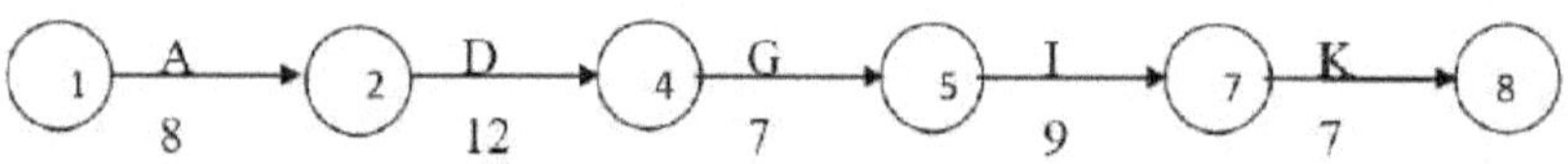

Time of the path = 8 + 12 + 7 + 9 + 7 = 43 weeks.

Compare the times for the three paths. Maximum of {52, 58, 45, 46, 33, 56, 43} = 58.

We see that the maximum time of a path is 58 weeks.

Forward pass

Calculation of Earliest Time of Occurrence of Events

Node	Earliest Time of Occurrence of Node
1	0
2	Time for Node 1 + Time for Activity A = 0+8=8
3	Time for Node 2 + Time for Activity B = 8+13=21
4	Time for Node 2 + Time for Activity D = 8+12= 20
5	Max {Time for Node 2 + Time for Activity C, Time for Node 3 + Time for Activity F, Time for Node 4 + Time for Activity G} = Max { 8 + 9, 21 + 8 , 20 + 7 } = Max {17, 29, 27} = 29
6	Max {Time for Node 3 + Time for Activity E, Time for Node 5 + Time for Activity H} = Max {21 + 14 , 29 + 12} = Max {35, 41} = 41
7	Max {Time for Node 5 + Time for Activity I, Time for Node 6 + Time for Activity J} = Max {29 + 9, 41 + 10} = Max {38, 51} = 51
8	Time for Node 7 + Time for Activity J = 51+ 7 = 58

Earliest Start Times of the activities

Activity	Earliest Start Time (Weeks)
A	0
B	8

C	8
D	8
E	21
F	21
G	20
H	29
I	29
J	41
K51	

Backward pass

Calculation of Latest Allowable Time of Occurrence of Events

Node	Latest Allowable Time of Occurrence of Node
8	Maximum time of a path in the network = 58
7	Time for Node 8 - Time for Activity K = 58 -7 = 51
6	Time for Node 7 - Time for Activity J = 51 -10 = 41
5	Min {Time for Node 6 - Time for Activity H, Time for Node 7 - Time for Activity I} = Min {41 - 12, 51 - 9} = Min {29, 42} = 29
4	Time for Node 5 - Time for Activity G = 29 - 7 = 22
3	Min {Time for Node 5 - Time for Activity F, Time for Node 6 - Time for Activity E} = Min {29 - 8, 41 - 14} = Min {21, 27} = 21

2	Min {Time for Node 3 - Time for Activity B, Time for Node 4 - Time for Activity D, = Min {21 - 13, 22 - 12, 29 - 9} = Min {8, 10, 20} = 8
1	Time for Node 2 - Time for Activity A = 8 - 8 = 0

Latest Finish Times of the activities

Activity	**Latest Finish Time (Weeks)**
K	58
J	51
I	51
H	41
G	29
F	29
E	41
D	22
C	29
B	21
A 8	

Calculation of Total Float for each activity:

Activity	Duration (Weeks)	Earliest Start Time	Earliest Finish Time	Latest Start Time	Latest Finish Time	Total Float = Latest Finish Time - Earliest Finish Time
A	8	0	8	0	8	0
B	13	8	21	8	21	0
C	9	8	17	20	29	12
D	12	8	20	10	22	2
E	14	21	35	27	41	6
F	8	21	29	21	29	0
G	7	20	27	22	29	2
H	12	29	41	29	41	0
I	9	29	38	42	51	13
J	10	41	51	41	51	0
K	7	51	58	51	58	0

The activities with total float = 0 are A, B, F, H, J and K. They are the critical activities.

Project completion time = 58 weeks.

Calculation of slacks of the events

Slack of an event = Latest Allowable Time of Occurrence of the event - Earliest Expected Time of Occurrence of that event.

Event (Node)	Earliest Expected Time of Occurrence of Event	Latest Allowable Time of Occurrence of Event	Slack of the Event
1	0	0	0
2	8	8	0
3	21	21	0
4	20	22	2
5	29	29	0
6	41	41	0
7	51	51	0
8	58	58	0

Interpretation

On the basis of the slacks of the events, it is concluded that the occurrence of event 4 may be delayed upto a maximum period of 2 weeks while no other event cannot be delayed.

Questions

1. Explain the terms: The earliest and latest times of the activities of a project.
2. Explain the procedure to find the earliest expected time of an event.
3. Explain the procedure to find the latest allowable time of an event.
4. What is meant by the slack of an activity? How will you determine it?
5. Consider the project with the following details:

activity	Duration (weeks)
1→ 2	1
2 → 3	3
2→ 4	7
3→ 4	5
3→ 5	8
4→ 5	4
5→ 6	1

Determine the earliest and the latest times of the activities. Calculate the total float for each activity and the slacks of the events.

****

Crashing Of A Project

Lesson Outline

- The Idea Of Crashing Of A Project
- The Criterion Of Selection Of An Activity For Crashing
- Numerical Problems

Learning Objectives

After reading this lesson you should be able to

- understand the concept of crashing of a project
- choose an activity for crashing
- work out numerical problems

The Meaning Of Crashing

The process of shortening the time to complete a project is called crashing and is usually achieved by putting into service additional labour or machines to one activity or more activities. Crashing involves more costs. A project manager would like to speed up a project by spending as minimum extra cost as possible. Project crashing seeks to minimize the extra cost for completion of a project before the stipulated time.

Steps In Project Crashing

Assumption: It is assumed that there is a linear relationship between time and cost.

Let us consider project crashing by the critical path method. The following four-step procedure is adopted.

Step 1

Find the critical path with the normal times and normal costs for the activities and identify the critical activities.

Step 2

Find out the crash cost per unit time for each activity in the network. This is calculated by means of the following formula.

$$\text{Crash cost}/\text{Time period} = \frac{\text{Crash cost} - \text{Normal cost}}{\text{Normal time} - \text{Crash time}}$$

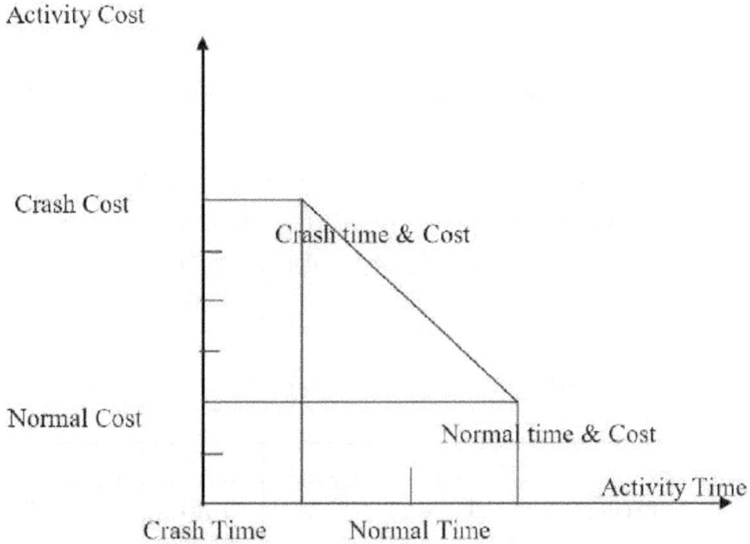

Step 3

Select an activity for crashing. The criteria for the selection is as follows:

Select the activity on the critical path with the smallest crash cost per unit time. Crash this activity to the maximum units of time as may be permissible by the given data.

Crashing an activity requires extra amount to be spent. However, even if the company is prepared to spend extra money, the activity time cannot be reduced beyond a certain limit in view of several other factors.

In step 1, we have to note that reducing the time of on activity along the critical path alone will reduce the completion time of a project. Because of this reason, we select an activity along the critical path for crashing. In step 3, we have to consider the following question:

If we want to reduce the project completion time by one unit, which critical activity will involve the least additional cost?

On the basis of the least additional cost, a critical activity is chosen for crashing. If there is a tie between two critical activities, the tie can be resolved arbitrarily.

Step 4

After crashing an activity, find out which is the critical path with the changed conditions. Sometimes, a reduction in the time of an activity in the critical path may cause a non-critical path to become critical. If the critical path with which we started is still the longest path, then go to Step 3. Otherwise, determine the new critical path and then go to Step 3.

Problem 1

A project has activities with the following normal and crash times and cost: .

Activity	Predecessor Activity	Normal Time (Weeks)	Crash Time (Weeks)	Normal Cost (Rs.)	Crash Cost (Rs.)
A	-	4	3	8,000	9,000
B	A	5	3	16,000	20,000
C	A	4	3	12,000	13,000
D	B	6	5	34,000	35,000
E	C	6	4	42,000	44,000
F	D	5	4	16,000	16,500
G	E	7	4	66,000	72,000
H	G	4	3	2,000	5,000

Determine a crashing scheme for the above project so that the total project time is reduced by 3 weeks.

Solution

We have the following network diagram for the given project with normal costs:

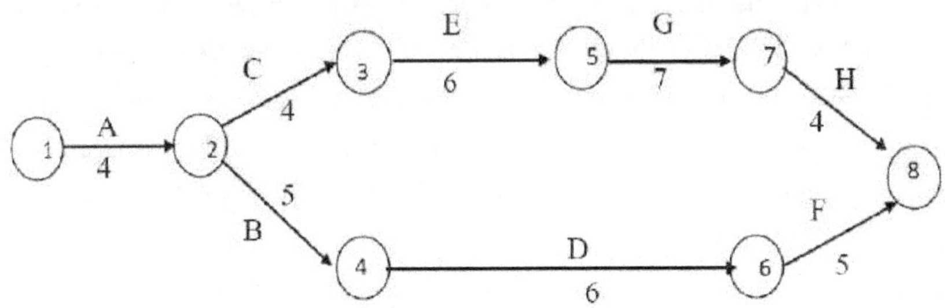

Beginning from the Start Node and terminating with the End Node, there are two paths for the network as detailed below:

Path I

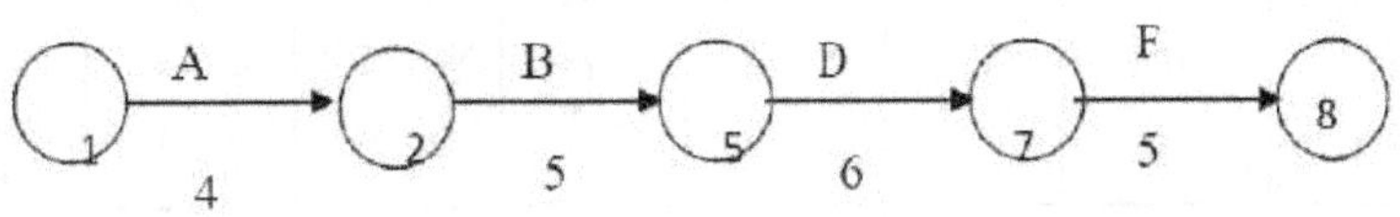

The time for the path = 4 + 5 + 6 + 5 = 20 weeks.

Path II

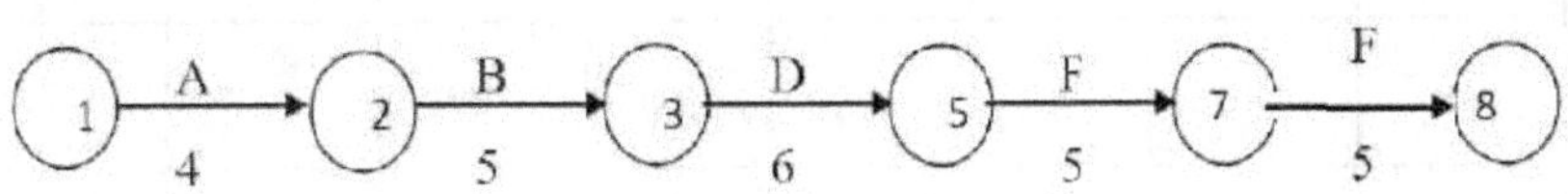

The time for the path = 4 + 4 + 6 + 7 + 4 = 25 weeks.
Maximum of {20, 25} = 25.

Therefore Path II is the critical path and the critical activities are A, C, E, G and H. The non-critical activities are B, D and F.

Given that the normal time of activity A is 4 weeks while its crash time is 3 weeks. Hence the time of this activity can be reduced by one week if the management is prepared to spend an additional amount. However, the time cannot be reduced by more than one week even if the management may be prepared to spend more money. The normal cost of this activity is Rs. 8,000 whereas the crash cost is Rs. 9,000. From this, we see that crashing of activity A by one week will cost the management an extra amount of Rs. 1,000. In a similar fashion, we can work out the crash cost per unit time for the other activities also. The results are provided in the following table.

Activity	Normal Time	Crash Time	Normal Cost	Crash Cost	Crash cost - Normal Cost	Normal Time - Crash Time	Crash Cost per unit time
A	4	3	8,000	9,000	1,000	1	1,000
B	5	3	16,000	20,000	4,000	2	2,000
C	4	3	12,000	13,000	1,000	1	1,000
D	6	5	34,000	35,000	1,000	1	1,000
E	6	4	42,000	44,000	2,000	2	1,000
F	5	4	16,000	16,500	500	1	500
G	7	4	66,000	72,000	6,000	1	6,000
H	4	3	2,000	5,000	3,000	1	3,000

A non-critical activity can be delayed without delaying the execution of the whole project. But, if a critical activity is delayed, it will delay the whole project. Because of this reason, we have to select a critical activity for crashing. Here we have to choose one of the activities A, C, E, G and H The crash cost per unit time works out as follows:

Rs. 1,000 for A; Rs. 1,000 for C; Rs. 1,000 for E; Rs. 6,000 for G; Rs. 3,000 for H.

The maximum among them is Rs. 1,000. So we have to choose an activity with Rs.1,000 as the crash cost per unit time. However, there is a tie among A, C and E. The tie can be resolved arbitrarily. Let us select A for crashing. We reduce the time of A by one week by spending an extra amount of Rs. 1,000.

After this step, we have the following network with the revised times for the activities:

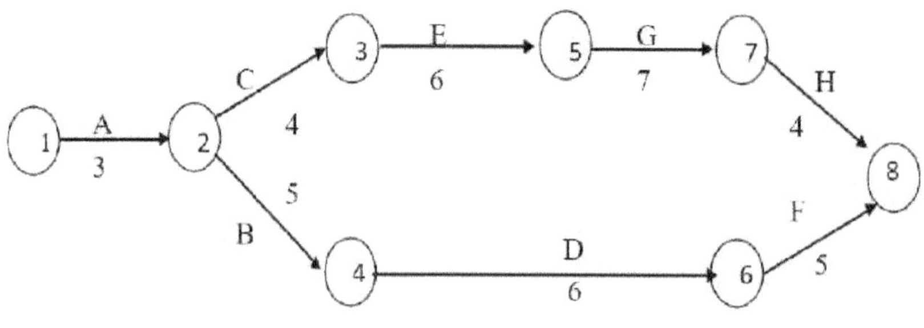

The revised time for Path I = 3 + 5 + 6 + 5 = 19 weeks.
The time for Path II = 3 + 4 + 6 + 7 + 4 = 24 weeks.
Maximum of {19, 24} = 24.

Therefore Path II is the critical path and the critical activities are A, C, E, G and H. However, the time for A cannot be reduced further. Therefore, we have to consider C, E, G and H for crashing. Among them, C and E have the least crash cost per unit time. The tie between C and E can be resolved arbitrarily. Suppose we reduce the time of C by one week with an extra cost of Rs. 1,000.

After this step, we have the following network with the revised times for the activities:

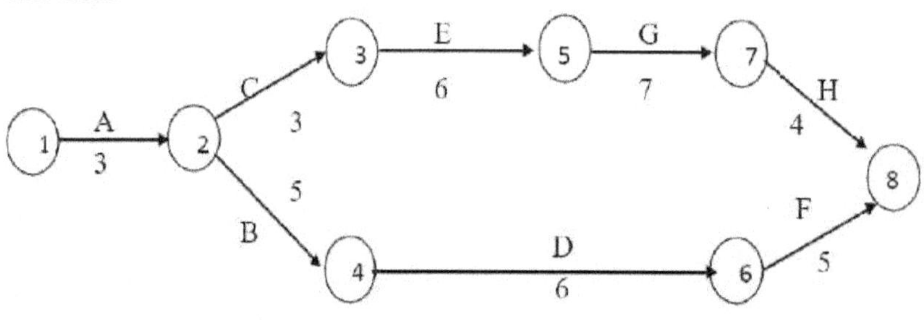

The time for Path I = 3 + 5 + 6 + 5 = 19 weeks.

The time for Path II = 3 + 3 + 6 + 7 + 4 = 23 weeks.

Maximum of {19, 23} = 23.

Therefore Path II is the critical path and the critical activities are A, C, E, G and H. Now the time for A or C cannot be reduced further. Therefore, we have to consider E, G and H for crashing. Among them, E has the least crash cost per unit time. Hence we reduce the time of E by one week with an extra cost of Rs. 1,000.

By the given condition, we have to reduce the project time by 3 weeks. Since this has been accomplished, we stop with this step.

Result: We have arrived at the following crashing scheme for the given project:

Reduce the time of A, C and E by one week each.

Project time after crashing is 22 weeks.

Extra amount required = 1,000 + 1,000 + 1,000 = Rs. 3,000.

Problem 2

The management of a company is interested in crashing of the following project by spending an additional amount not exceeding Rs. 2,000. Suggest how this can be accomplished.

Activity	Predecessor Activity	Normal Time (Weeks)	Crash Time (Weeks)	Normal Cost (Rs.)	Crash Cost (Rs.)
A	-	7	6	15,000	18,000
B	A	12	9	11,000	14,000
C	A	22	21	18,500	19,000
D	B	11	10	8,000	9,000
E	C, D	6	5	4,000	4,500

Solution

We have the following network diagram for the given project with normal costs:

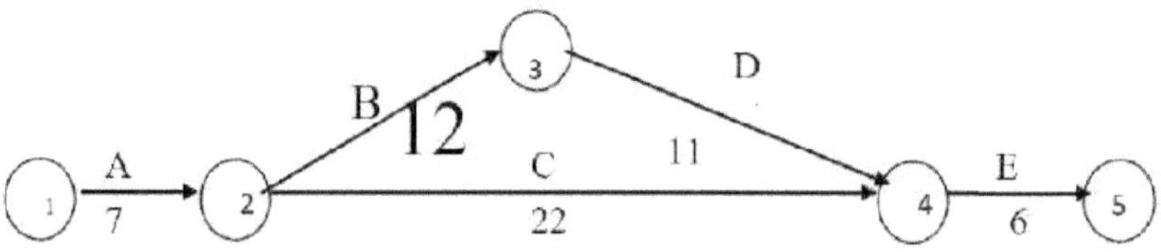

There are two paths for this project as detailed below:

Path I

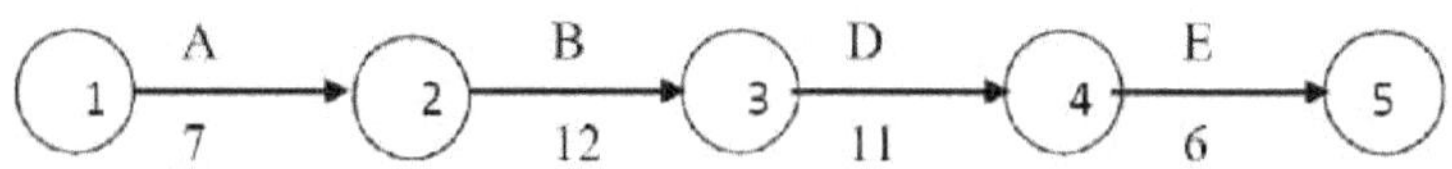

The time for the path = 7 + 12 + 11 + 6 = 36 weeks.

Path II

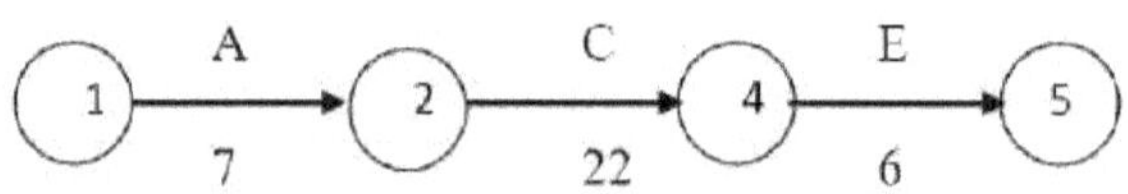

The time for the path = 7 + 22 + 6 = 35 weeks.

Maximum of {36, 35} = 36.

Therefore Path I is the critical path and the critical activities are A, B, D and E. The non-critical activity is C.

The crash cost per unit time for the activities in the project are provided in the following table.

Activity	Normal	Crash Time	Normal Cost	Crash Cost	Crash cost - Normal Cost	Normal Time - Crash Time	Crash Cost per unit time
A	7	6	15,000	18,000	3,000	1	3,000
B	12	9	11,000	14,000	3,000	3	1,000
C	22	21	18,500	19,000	500	1	500
D	11	10	8,000	9,000	1,000	1	1,000
E	6	5	4,000	4,500	500	1	500

We have to choose one of the activities A, B, D and E for crashing. The crash cost per unit time is as follows:

Rs. 3,000 for A; Rs. 1,000 for B; Rs. 1,000 for D; Rs. 500 for E.

The least among them is Rs. 500. So we have to choose the activity E for crashing. We reduce the time of E by one week by spending an extra amount of Rs. 500.

After this step, we have the following network with the revised times for the activities:

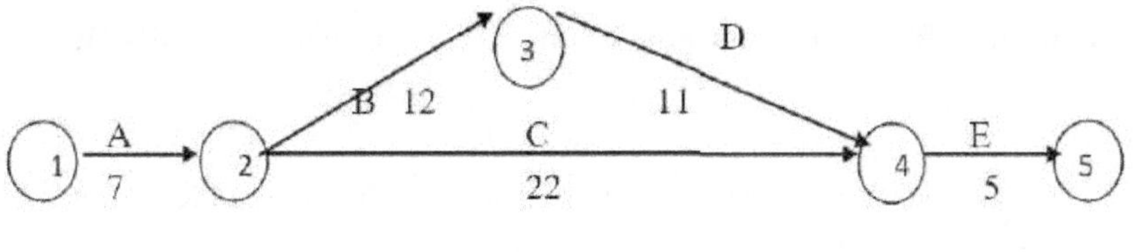

The revised time for Path I = 7 + 12 + 11 + 5 = 35 weeks.
The time for Path II = 7 + 22 + 5 = 34 weeks.
Maximum of {35, 34} = 35.

Therefore Path I is the critical path and the critical activities are A, B, D and E. The non-critical activity is C.

The time of E cannot be reduced further. So we cannot select it for crashing. Next B and have the smallest crash cost per unit time. Let us select B for crashing. Let us reduce the time of E by one week at an extra cost of Rs. 1,000.

After this step, we have the following network with the revised times for the activities:

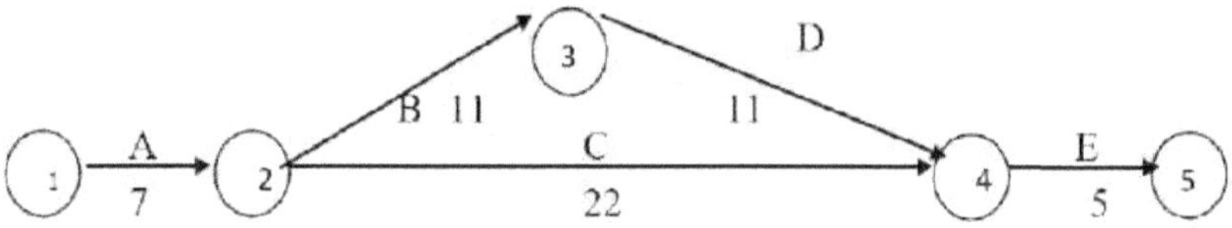

The revised time for Path I = 7 + 11 + 11 + 5 = 34 weeks.
The time for Path II = 7 + 22 + 5 = 34 weeks.
Maximum of {34, 34} = 34.

Since both paths have equal times, both are critical paths. So, we can choose an activity for crashing from either of them depending on the least crash cost per unit time. In path I, the activities are A, B, D and E. In path II, the activities are A, C and E.

The crash cost per unit time is the least for activity C. So we select C for crashing. Reduce the time of C by one week at an extra cost of Rs. 500.

By the given condition, the extra amount cannot exceed Rs. 2,000. Since this state has been met, we stop with this step.

Result: The following crashing scheme is suggested for the given project:

Reduce the time of E, B and C by one week each.

Project time after crashing is 33 weeks.

Extra amount required = 500 + 1,000 + 500 = Rs. 2,000.

Problem 3

The manager of a company wants to apply crashing for the following project by spending an additional amount not exceeding Rs. 2,000. Offer your suggestion to the manager.

Activity	Predecessor Activity	Normal Time (Weeks)	Crash Time (Weeks)	Normal Cost (Rs.)	Crash Cost (Rs.)
A	-	20	19	8,000	10,000
B	-	15	14	16,000	19,000
C	A	22	20	13,000	14,000
D	A	17	15	7,500	9,000
E	B	19	18	4,000	5,000
F	C	28	27	3,000	4,000
G	D, E	25	24	12,000	13,000

Solution

We have the following network diagram for the given project with normal costs:

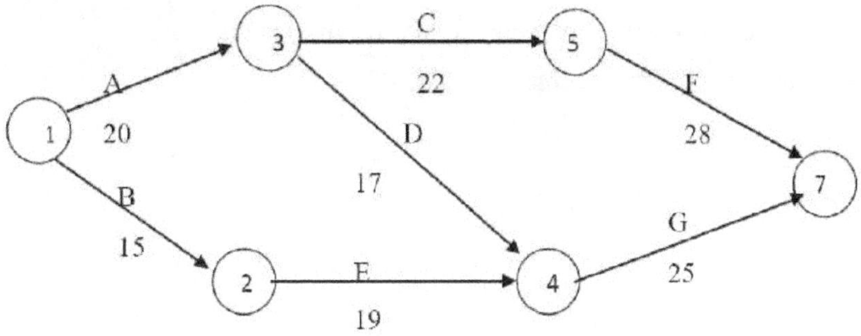

There are three paths for this project as detailed below:

Path I

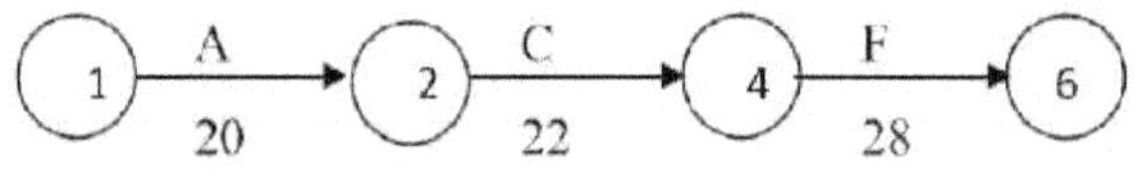

The time for the path = 20 + 22 + 28= 70 weeks.

Path II

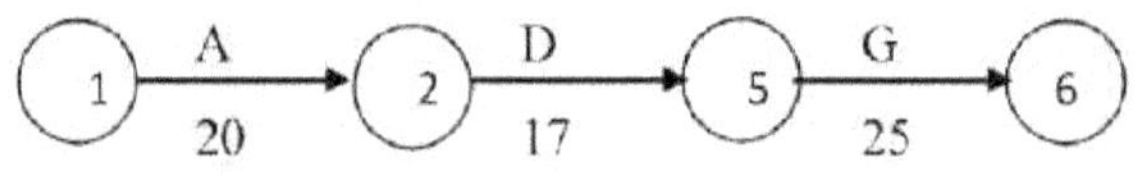

The time for the path = 20 + 17 + 25= 62 weeks.

Path III

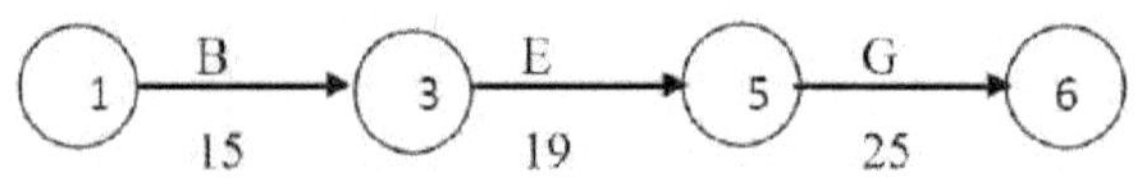

The time for the path = 15+19 +25 = 69 weeks.
Maximum of {70, 62, 69} = 70.

Therefore Path I is the critical path and the critical activities are A, C and F. The non-critical activities are B, D, E and G.

The crash cost per unit time for the activities in the project are provided in the following table

Activity	Normal Time	Crash Time	Normal Cost	Crash Cost	Crash cost - Normal Cost	Normal Time - Crash Time	Crash Cost per unit time
A	20	19	8,000	10,000	2,000	1	2,000
B	15	14	16,000	19,000	3,000	1	3,000
C	22	20	13,000	14,000	1,000	2	500
D	17	15	7,500	9,000	1,500	2	750
E	19	18	4,000	5,000	1,000	1	1,000
F	28	27	3,000	4,000	1,000	1	1,000
G	25	24	12,000	13,000	1,000	1	1,000

We have to choose one of the activities A, C and F for crashing. The crash cost per unit time is as follows:

Rs. 2,000 for A; Rs. 500 for C; Rs. 1,000 for F.

The least among them is Rs. 500. So we have to choose the activity C for crashing. We reduce the time of C by one week by spending an extra amount of Rs. 500.

After this step, we have the following network with the revised times for the activities:

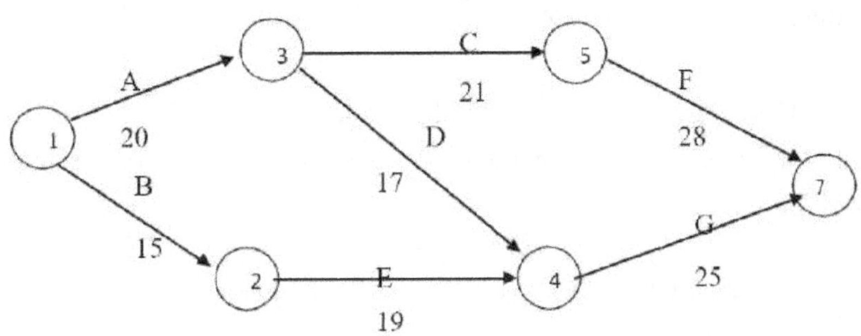

The revised time for Path I = 20 + 21 + 28= 69 weeks.
The time for Path II = 20 + 17 + 25= 62 weeks.
The time for Path III = 15+19 +25 = 69 weeks.
Maximum of {69, 62, 69} = 69.

Since paths I and III have equal times, both are critical paths. So, we can choose an activity for crashing from either of them depending on the least crash cost per unit time.

In path I, the activities are A, C and F. In path III, the activities are B, E and G.

The crash cost per unit time is the least for activity C. So we select C for crashing. Reduce the time of C by one week at an extra cost of Rs. 500.

After this step, we have the following network with the revised times for the activities:

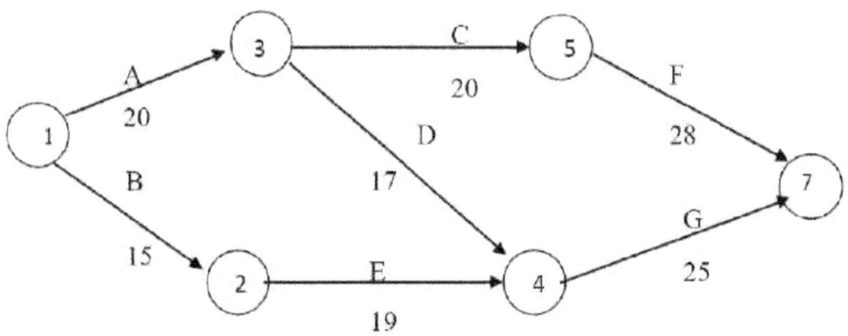

The revised time for Path I = 20 + 20 + 28= 68 weeks.
The time for Path II = 20 + 17 + 25= 62 weeks.
The time for Path III = 15+19 +25 = 69 weeks.
Maximum of {68, 62, 69} = 69.

Therefore path III is the critical activities. Hence we have to select an activity from Path III for crashing. We see that the crash cost per unit time is as follows:

Rs. 3,000 for B; Rs. 1,000 for E; Rs. 1,000 for G.

The least among them is Rs. 1,000. So we can select either E or G for crashing. Let us select E for crashing. We reduce the time of E by one week by spending an extra amount of Rs. 1,000.

By the given condition, the extra amount cannot exceed Rs. 2,000. Since this condition has been reached, we stop with this step.

Result: The following crashing scheme is suggested for the given project:
Reduce the time of C by 2 weeks and that of E by one week.

Project time after crashing is 67 weeks.
Extra amount required = 2 x 500 + 1,000 = Rs. 2,000.

Questions

1. Explain the concept of crashing of a project.
2. Explain the criterion for selection of an activity for crashing.

****

Game Theory, Goal Programming & Queuing Theory

Lesson 1 - Basic Concepts In Game Theory

Lesson Outline

- Introduction To The Theory Of Games
- The Definition Of A Game
- Competitive Game
- Managerial Applications Of The Theory Of Games
- Key Concepts In The Theory Of Games
- Types Of Games

Learning Objectives

After reading this lesson you should be able to

- Understand the concept of a game
- Grasp the assumptions in the theory of games
- Appreciate the managerial applications of the theory of games
- Understand the key concepts in the theory of games
- Distinguish between different types of games

Introduction To Game Theory

Game theory seeks to analyse competing situations which arise out of conflicts of interest. Abraham Maslow's hierarchical model of human needs lays emphasis on fulfilling the basic needs such as food, water, clothes, shelter, air, safety and security. There is conflict of interest between animals and plants in the consumption of natural resources. Animals compete among themselves for securing food. Man competes with animals to earn his food. A man also competes with another man. In the past, nations waged wars to expand the territory of their rule. In the present day world, business organizations compete with each other in getting the market share. The conflicts of interests of human beings are not confined to the basic needs alone. Again considering Abraham Maslow's model of human needs, one can realize that conflicts also arise due to the higher levels of human needs such as love, affection, affiliation, recognition, status, dominance, power, esteem, ego, self-respect, etc. Sometimes one witnesses clashes of ideas of intellectuals also. Every intelligent and rational participant in a conflict wants to be a winner but not all participants can be the winners at a time. The situations of conflict gave birth to Darwin's theory of the 'survival of the fittest'. Nowadays the concepts of conciliation, co-existence, co-operation, coalition and consensus are gaining ground. Game theory is another tool to examine situations of conflict so as to identify the courses of action to be followed and to take appropriate decisions in the long run. Thus this theory assumes importance from managerial perspectives. The pioneering work on the theory of games was done by von Neumann and Morgenstern through their publication entitled 'The Theory of Games and Economic Behaviour' and subsequently the subject was developed by several experts. This theory can offer valuable guidelines to a manager in 'strategic management' which can be used in the decision making process for merger, take-over, joint venture, etc. The results obtained by the application of this theory can serve as an early warning to the top level management in meeting the threats from the competing business organizations and for the conversion of the internal weaknesses and external threats into opportunities and strengths, thereby achieving the goal of maximization of profits. While this theory does not describe any procedure to play a game, it will enable a participant to select the appropriate strategies to be followed in the pursuit of his goals. The situation of failure in a game would activate a participant in the analysis of the relevance of the existing strategies and lead him to identify better, novel strategies for the future occasions.

Definitions of game theory

There are several definitions of game theory. A few standard definitions are presented below.

In the perception of Robert Mockler, "Game theory is a mathematical technique helpful in making decisions in situations of conflicts, where the success of one part depends at the expense of others, and where the individual decision maker is not in complete control of the factors influencing the outcome".

The definition given by William G. Nelson runs as follows: "Game theory, more properly **the theory of games of strategy,** is a mathematical method of analyzing a conflict. The alternative is not between this decision or that decision, but between this strategy or that strategy to be used against the conflicting interest".

In the opinion of Matrin Shubik, "Game theory is a method of the study of decision making in situation of conflict. It deals with human processes in which the individual decision-unit is not in complete control of other decision-units entering into the environment".

According to von Neumann and Morgenstern, "The 'Game' is simply the totality of the rules which describe it. Every particular instance at which the game is played – in a particular way – from beginning to end is a 'play'. The game consists of a sequence of moves, and the play of a sequence of choices".

J.C.C McKinsey points out a valid distinction between two words, namely 'game' and 'play'. According to him, "game refers to a particular realization of the rules".

In the words of O.T. Bartos, "The theory of games can be used for 'prescribing' how an intelligent person should go about resolving social conflicts, ranging all the way from open warfare between nations to disagreements between husband and wife".

Martin K Starr gave the following definition: "Management models in the competitive sphere are usually termed game models. By studying game theory, we can obtain substantial information into management's role under competitive conditions, even though much of the game theory is neither directly operational nor implementable".

According to Edwin Mansfield, "A game is a competitive situation where two or more persons pursue their own interests and no person can dictate the outcome. Each player, an entity with the same interests, make his own decisions. A player can be an individual or a group".

Assumptions for a Competitive Game

Game theory helps in finding out the best course of action for a firm in view of the anticipated countermoves from the competing organizations. A competitive situation is a competitive game if the following properties hold:

The number of competitors is finite, say N.

A finite set of possible courses of action is available to each of the N competitors.

A play of the game results when each competitor selects a course of action from the set of courses available to him. In game theory we make an important assumption that al the players select their courses of action simultaneously. As a result, no competitor will be in a position to know the choices of his competitors.

The outcome of a play consists of the particular courses of action chosen by the individual players. Each outcome leads to a set of payments, one to each player, which may be either positive, or negative, or zero.

Managerial Applications of the Theory of Games

The techniques of game theory can be effectively applied to various managerial problems as detailed below:

1. Analysis of the market strategies of a business organization in the long run.
2. Evaluation of the responses of the consumers to a new product.
3. Resolving the conflict between two groups in a business organization.
4. Decision making on the techniques to increase market share.
5. Material procurement process.
6. Decision making for transportation problem.
7. Evaluation of the distribution system.
8. Evaluation of the location of the facilities.
9. Examination of new business ventures and
10. Competitive economic environment.

Key concepts in the Theory of Games

Several of the key concepts used in the theory of games are described below:

Players

The competitors or decision makers in a game are called the players of the game.

Strategies

The alternative courses of action available to a player are referred to as his strategies.

Pay off

The outcome of playing a game is called the pay off to the concerned player.

Optimal Strategy

A strategy by which a player can achieve the best pay off is called the optimal strategy for him.

Zero-sum game

A game in which the total payoffs to all the players at the end of the game is zero is referred to as a zero-sum game.

Non-zero sum game

Games with "less than complete conflict of interest" are called non-zero sum games. The problems faced by a large number of business organizations come under this category. In such games, the gain of one player in terms of his success need not be completely at the expense of the other player.

Payoff matrix

The tabular display of the payoffs to players under various alternatives is called the payoff matrix of the game.

Pure strategy

If the game is such that each player can identify one and only one strategy as the optimal strategy in each play of the game, then that strategy is referred to as the best strategy for that player and the game is referred to as a game of pure strategy or a pure game.

Mixed strategy

If there is no one specific strategy as the 'best strategy' for any player in a game, then the game is referred to as a game of mixed strategy or a mixed game. In such a game, each player has to choose different alternative courses of action from time to time.

N-person game

A game in which N-players take part is called an N-person game.

Maximin-Minimax Principle

The maximum of the minimum gains is called the maximin value of the game and the corresponding strategy is called the maximin strategy. Similarly the minimum of the maximum losses is called the minimax value of the game and the corresponding strategy is called the minimax strategy. If both the values are equal, then that would guarantee the best of the worst results.

Negotiable or cooperative game

If the game is such that the players are taken to cooperate on any or every action which may increase the payoff of either player, then we call it a negotiable or cooperative game.

Non-negotiable or non-cooperative game

If the players are not permitted for coalition then we refer to the game as a non-negotiable or non-cooperative game.

Saddle point

A saddle point of a game is that place in the payoff matrix where the maximum of the row minima is equal to the minimum of the column maxima. The payoff at the saddle point is called the value of the game and the corresponding strategies are called the pure strategies.

Dominance

One of the strategies of either player may be inferior to at least one of the remaining ones. The superior strategies are said to dominate the inferior ones.

Types of Games

There are several classifications of a game. The classification may be based on various factors such as the number of participants, the gain or loss to each participant, the number of strategies available to each participant, etc. Some of the important types of games are enumerated below.

Two person games and n-person games

In two person games, there are exactly two players and each competitor will have a finite number of strategies. If the number of players in a game exceeds two, then we refer to the game as n-person game.

Zero sum game and non-zero sum game

If the sum of the payments to all the players in a game is zero for every possible outcome of the game, then we refer to the game as a zero sum game. If the sum of the payoffs from any play of the game is either positive or negative but not zero, then the game is called a non-zero sum game

Games of perfect information and games of imperfect information

A game of perfect information is the one in which each player can find out the strategy that would be followed by his opponent. On the other hand, a game of imperfect information is the one in which no player can know in advance what strategy would be adopted by the competitor and a player has to proceed in his game with his guess works only.

Games with finite number of moves / players and games with unlimited number of moves

A game with a finite number of moves is the one in which the number of moves for each player is limited before the start of the play. On the other hand, if the game can be continued over an extended period of time and the number of moves for any player has no restriction, then we call it a game with unlimited number of moves.

Constant-sum games

If the sum of the game is not zero but the sum of the payoffs to both players in each case is constant, then we call it a constant sum game. It is possible to reduce such a game to a zero-sum game.

2x2 two person game and 2xn and mx2 games

When the number of players in a game is two and each player has exactly two strategies, the game is referred to as 2x2 two person game.

A game in which the first player has precisely two strategies and the second player has three or more strategies is called an 2xn game.

A game in which the first player has three or more strategies and the second player has exactly two strategies is called an mx2 game.

3x3 and large games

When the number of players in a game is two and each player has exactly three strategies, we call it a 3x3 two person game.

Two-person zero sum games are said to be larger if each of the two players has 3 or more choices.

The examination of 3x3 and larger games is involves difficulties. For such games, the technique of linear programming can be used as a method of solution to identify the optimum strategies for the two players.

Non-constant games

Consider a game with two players. If the sum of the payoffs to the two players is not constant in all the plays of the game, then we call it a non-constant game.

Such games are divided into negotiable or cooperative games and non-negotiable or non-cooperative games.

Questions

1. Explain the concept of a game.
2. Define a game.
3. State the assumptions for a competitive game.
4. State the managerial applications of the theory of games.
5. Explain the following terms: strategy, pay-off matrix, saddle point, pure strategy and mixed strategy.
6. Explain the following terms: two person game, two person zero sum game, value of a game, 2xn game and mx2 game.

****

Lesson 2 - Two-Person Zero Sum Games

Lesson Outline

- The Concept Of A Two-Person Zero Sum Game
- The Assumptions For A Two-Person Zero Sum Game
- Minimax And Maximin Principles

Learning Objectives

After reading this lesson you should be able to

- Understand the concept of a two-person zero sum game
- Have an idea of the assumptions for a two-person zero sum game
- Understand Minimax and Maximin principles
- Solve a two-person zero sum game
- Interpret the results from the payoff matrix of a two-person zero sum game

Definition of two-person zero sum game

A game with only two players, say player A and player B, is called a two-person zero sum game if the gain of the player A is equal to the loss of the player B, so that the total sum is zero.

Payoff matrix

When players select their particular strategies, the payoffs (gains or losses) can be represented in the form of a payoff matrix.

Since the game is zero sum, the gain of one player is equal to the loss of other and vice-versa. Suppose A has m strategies and B has n strategies. Consider the following payoff matrix.

Player B's strategies

Player A's strategies	B_1	B_2	B_n
A_1	a_{11}	a_{12}	a_{1n}
A_2	a_{21}	a_{22}	a_{2n}
A_m	a_{m1}	a_{m2}	a_{mn}

Player A wishes to gain as large a payoff a_{ij} as possible while player B will do his best to reach as small a value a_{ij} as possible where the gain to player B and loss to player A be ($- a_{ij}$).

Assumptions for two-person zero sum game

For building any model, certain reasonable assumptions are quite necessary. Some assumptions for building a model of two-person zero sum game are listed below.

a. Each player has available to him a finite number of possible courses of action. Sometimes the set of courses of action may be the same for each player. Or, certain courses of action may be available to both players while each player may have certain specific courses of action which are not available to the other player.
b. Player A attempts to maximize gains to himself. Player B tries to minimize losses to himself.
c. The decisions of both players are made individually prior to the play with no communication between them.
d. The decisions are made and announced simultaneously so that neither player has an advantage resulting from direct knowledge of the other player's decision.
e. Both players know the possible payoffs of themselves and their opponents.

Minimax and Maximin Principles

The selection of an optimal strategy by each player without the knowledge of the competitor's strategy is the basic problem of playing games.

The objective of game theory is to know how these players must select their respective strategies, so that they may optimize their payoffs. Such a criterion of decision making is referred to as minimax-maximin principle. This principle in games of pure strategies leads to the best possible selection of a strategy for both players.

For example, if player A chooses his i_{th} strategy, then he gains at least the payoff min a_{ij} , which is minimum of the i_{th} row elements in the payoff matrix. Since his objective is to maximize his payoff, he can choose strategy i so as to make his payoff as large as possible. i.e., a payoff which is not less than $\max\limits_{1 \le i \le m} \min\limits_{1 \le j \le n} a_{ij}$.

Similarly player B can choose jth column elements so as to make his loss not greater than $\max\limits_{1 \le i \le m} \min\limits_{1 \le j \le n} a_{ij}$.

If the maximin value for a player is equal to the minimax value for another player, i.e.

$$\max_{1 \le i \le m} \min_{1 \le j \le n} a_{ij} = V = \min_{1 \le j \le n} \max_{1 \le i \le m} a_{ij}$$

then the game is said to have a saddle point (equilibrium point) and the corresponding strategies are called optimal strategies. If there are two or more saddle points, they must be equal.

The amount of payoff, i.e., V at an equilibrium point is known as the ***value of the game.***

The optimal strategies can be identified by the players in the long run.

Fair game

The game is said to be fair if the value of the game V = 0.

Problem 1

Solve the game with the following pay-off matrix.

Player B

Strategies

Player A Strategies		*I*	*II*	*III*	*IV*	*V*
	1	−2	5	− 3	6	7
	2	4	6	8	−1	6
	3	8	2	3	5	4
	4	15	14	18	12	20

Solution

First consider the minimum of each row.

Row	Minimum Value
1	-3
2	-1
3	2
4	12

Maximum of {-3, -1, 2, 12} = 12

Next consider the maximum of each column.

Column	Maximum Value
1	15
2	14
3	18
4	12
5	20

Minimum of {15, 14, 18, 12, 20}= 12

We see that the maximum of row minima = the minimum of the column maxima. So the game has a saddle point. The common value is 12. Therefore the value V of the game = 12.

Interpretation

In the long run, the following best strategies will be identified by the two players:

The best strategy for player A is strategy 4.
The best strategy for player B is strategy IV.
The game is favourable to player A.

Problem 2

Solve the game with the following pay-off matrix

Player Y
Strategies

Player X Strategies	*I*	*II*	*III*	*IV*	*V*
1	9	12	7	14	26
2	25	35	20	28	30
3	7	6	−8	3	2
4	8	11	13	−2	1

Solution

First consider the minimum of each row.

Row	Minimum Value
1	7
2	20
3	-8
4	-2

Maximum of {7, 20, −8, -2} = 20

Next consider the maximum of each column.

Column	Maximum Value
1	25
2	35
3	20
4	28
5	30

Minimum of {25, 35, 20, 28, 30}= 20

It is observed that the maximum of row minima and the minimum of the column maxima are equal. Hence the given the game has a saddle point. The common value is 20. This indicates that the value V of the game is 20.

Interpretation.

The best strategy for player X is strategy 2.

The best strategy for player Y is strategy III.

The game is favourable to player A.

Problem 3

Solve the following game:

Player B

Strategies

Player A Strategies		I	II	III	IV
	1	1	−6	8	4
	2	3	−7	2	−8
	3	5	−5	−1	0
	4	3	−4	5	7

Solution

First consider the minimum of each row.

Row	Minimum Value
1	-6
2	-8
3	-5
4	-4

Maximum of {-6, -8, -5, -4} = -4

Next consider the maximum of each column.

Column	Maximum Value
1	5
2	-4
3	8
4	7

Minimum of {5, -4, 8, 7}= - 4

Since the max {row minima} = min {column maxima}, the game under consideration has a saddle point. The common value is –4. Hence the value of the game is –4.

Interpretation

The best strategy for player A is strategy 4.

The best strategy for player B is strategy II.

Since the value of the game is negative, it is concluded that the game is favourable to player B.

Questions

1. What is meant by a two-person zero sum game? Explain.
2. State the assumptions for a two-person zero sum game.
3. Explain Minimax and Maximin principles.
4. How will you interpret the results from the payoff matrix of a two-person zero sum game? Explain.
5. What is a fair game? Explain.
6. Solve the game with the following pay-off matrix.

Player B

Strategies

Player A Strategies		*I*	*II*	*III*	*IV*	*V*
	1	7	5	2	3	9
	2	10	8	7	4	5
	3	9	12	0	2	1
	4	11	− 2	−1	3	4

Answer: Best strategy for A: 2

Best strategy for B: IV

V = 4

The game is favourable to player A

7. Solve the game with the following pay-off matrix.

Player B

Strategies

Player A Strategies		*I*	*II*	*III*	*IV*	*V*
	1	−2	− 3	8	7	0
	2	1	− 7	−5	−2	3
	3	4	− 2	3	5	−1
	4	6	− 4	5	4	7

Answer: Best strategy for A: 3

Best strategy for B: II

V=-2

The game is favourable to player B

8. Solve the game with the following pay-off matrix.

Player B

Player A	6	4
	7	5

Answer: Best strategy for A: 2

Best strategy for B: II

V = 5

The game is favourable to player A

9. Solve the following game and interpret the result.

Player B

Strategies

		I	*II*	*III*	*IV*
Player A Strategies	1	−3	−7	1	3
	2	−1	2	− 3	1
	3	0	4	2	6
	4	−2	−1	− 5	1

Answer: Best strategy for A: 3

Best strategy for B: I

V = 0

The value V = 0 indicates that the game is a fair one.

10. Solve the following game:

Player B

Strategies

		I	*II*	*III*
Player A Strategies	1	1	8	2
	2	3	5	6
	3	2	2	1

Answer: Best strategy for A: 2

Best strategy for B: I

V = 3

The game is favourable to player A

11. Solve the game

Player B

Player A

	I	II	III	IV
1	4	−1	2	0
2	−3	−5	−9	− 2
3	2	−8	0	−11

Answer : V = -1

12. Solve the game

Player Y

Player X

	I	II	III	IV	V
1	4	0	1	7	−1
2	0	−3	−5	−7	5
3	3	2	3	4	3
4	−6	4	−1	0	5
5	0	0	6	0	0

Answer : V = 2

13. Solve the game

Player B

Player A

	I	II	III	IV	V
1	9	3	4	4	2
2	8	6	8	5	12
3	10	7	19	18	14
4	8	6	8	11	6
5	3	5	16	10	8

Answer : V = 7

14. Solve the game

Player Y

Player X

17	10	12	5	4	8
2	5	6	7	6	9
7	6	9	2	3	1
10	11	14	8	13	8
20	18	17	10	15	17
12	11	15	9	5	11

Answer : V = 10

15. Solve the game

Player B

Player A

12	14	8	7	4	9
2	13	6	7	9	8
13	6	8	6	3	1
14	9	10	8	9	6
20	18	17	11	14	16
8	12	16	9	6	13

Answer : V = 11

16. Examine whether the following game is fair.

Player Y

Player X

6	−4	−3	−2
3	5	0	8
7	−2	−6	5

Answer : V = 0.

Therefore, it is a fair game.

****

Games With No Saddle Point

Lesson Outline

- The Concept Of A 2X2 Game With No Saddle Point
- The Method Of Solution

Learning Objectives

After reading this lesson you should be able to

- Understand the concept of a 2x2 game with no saddle point
- Know the method of solution of a 2x2 game without saddle point
- Solve a game with a given payoff matrix
- Interpret the results obtained from the payoff matrix

2 x 2 zero-sum game

When each one of the first player A and the second player B has exactly two strategies, we have a 2 x 2 game.

Motivating point

First let us consider an illustrative example

Problem 1

Examine whether the following 2 x 2 game has a saddle point

	Player B	
Player A	3	5
	4	2

Solution

First consider the minimum of each row.

Row	Minimum Value
1	3
2	2

Maximum of {3, 2} = 3

Next consider the maximum of each column.

Column	Maximum Value
1	4
2	5

Minimum of {4, 5}= 4

We see that max {row minima} and min {column maxima} are not equal. Hence the game has no saddle point.

Method of solution of a 2x2 zero-sum game without saddle point

Suppose that a 2x2 game has no saddle point. Suppose the game has the following pay-off matrix.

Player B

Player A $\begin{bmatrix} a & b \\ c & d \end{bmatrix}$

Since this game has no saddle point, the following condition shall hold:

$$Max\ \{Min\ \{a\ ,b\},Min\ \{c\ ,d\ \}\} \neq Min\ \{Max\ \{a\ ,c\},Max\ \{b,d\}\}$$

In this case, the game is called a mixed game. ***No strategy of Player A can be called the best strategy for him. Therefore A has to use both of his strategies. Similarly no strategy of Player B can be called the best strategy for him and he has to use both of his strategies.***

Let p be the probability that Player A will use his first strategy. Then the probability that Player A will use his second strategy is 1-p.

If Player B follows his first strategy

Expected value of the pay-off to Player A

= {Expected value of the pay-off to Player A arising from his first strategy} + {Expected value of the pay-off to Player A arising from his second strategy}

$$= ap + c\,(1-p) \qquad (1)$$

In the above equation, note that the expected value is got as the product of the corresponding values of the pay-off and the probability.

If Player B follows his second strategy

Expected value of the pay-off to Player A $= bp + d\,(1-p)$ (2)

If the expected values in equations (1) and (2) are different, Player B will prefer the minimum of the two expected values that he has to give to player A. Thus B will have a pure strategy. This contradicts our assumption that the game is a mixed one. Therefore the expected values of the pay-offs to Player A in equations (1) and (2) should be equal. Thus we have the condition

$$
\begin{aligned}
ap + c(1 - p) &= bp + d(1 - p) \\
ap - bp &= (1 - p)[d - c] \\
p(a - b) &= (d - c) - p(d - c) \\
p(a - b) + p(d - c) &= d - c \\
p(a - b + d - c) &= d - c \\
p &= \frac{d - c}{(a + d) - (b + c)} \\
1 - p &= \frac{a + d - b - c - d + c}{(a + d) - (b + c)} \\
&= \frac{a - b}{(a + d) - (b + c)}
\end{aligned}
$$

{The number of times A will use first strategy} : {The number of times A will use second strategy} = $\frac{d - c}{(a + d) - (b + c)} : \frac{a - b}{(a + d) - (b + c)}$

The expected pay-off to Player A

$$
\begin{aligned}
&= ap + c(1 - p) \\
&= c + p(a - c) \\
&= c + \frac{(d - c)(a - c)}{(a + d) - (b + c)} \\
&= \frac{c\{(a + d) - (b + c)\} + (d - c)(a - c)}{(a + d) - (b + c)} \\
&= \frac{ac + cd - bc - c^2 + ad - cd - ac + c^2)}{(a + d) - (b + c)} \\
&= \frac{ad - bc}{+ \quad d) - (b + c)(a}
\end{aligned}
$$

Therefore, the value V of the game is

$$\frac{ad - bc}{(a + d) - (b + c)}$$

To find the number of times that B will use his first strategy and second strategy:

Let the probability that B will use his first strategy be r. Then the probability that B will use his second strategy is 1-r.

When A use his first strategy

The expected value of loss to Player B with his first strategy = ar

The expected value of loss to Player B with his second strategy = b(1-r)

Therefore the expected value of loss to B = ar + b(1-r) (3)

When A use his second strategy

The expected value of loss to Player B with his first strategy = cr

The expected value of loss to Player B with his second strategy = d(1-r)

Therefore the expected value of loss to B = cr + d(1-r) (4)

If the two expected values are different then it results in a pure game, which is a contradiction. Therefore the expected values of loss to Player B in equations (3) and (4) should be equal. Hence we have the condition

$$
\begin{aligned}
ar + b(1-r) &= cr + d(1-r) \\
ar + b - br &= cr + d - dr \\
ar - br - cr + dr &= d - b \\
r(a - b - c + d) &= d - b \\
r &= \frac{d-b}{a-b-c+d} \\
&= \frac{d-b}{(a+d)-(b+c)}
\end{aligned}
$$

Problem 2

Solve the following game

Y

X $\begin{matrix} 2 \\ 1 \end{matrix}$

Type here

Solution

First consider the row minima.

Row	Minimum Value
1	2
2	1

Maximum of {2, 1} = 2

Next consider the maximum of each column

Column	Maximum Value
1	4
2	5

Minimum of {4, 5}= 4

We see that

Max {row minima} ≠ min {column maxima}

So the game has no saddle point. Therefore it is a mixed game.

We have a = 2, b = 5, c = 4 and d = 1.

Let p be the probability that player X will use his first strategy. We have

$$p = \frac{d - c}{(a + d) - (b + c)}$$

$$= \frac{1 - 4}{(2+1)-(5+4)}$$

$$= \frac{-3}{3 - 9}$$

$$= \frac{-3}{-6}$$

$$= \frac{1}{2}$$

The probability that player X will use his second strategy is

$1\text{-p} = 1 - \frac{1}{2} = \frac{1}{2}$.

Value of the game $V = \frac{ad - bc}{(a+d)-(b+c)} = \frac{2-20}{3-9} = \frac{-18}{-6} = 3$.

Let r be the probability that Player Y will use his first strategy. Then the probability that Y will use his second strategy is (1-r). We have

$$r = \frac{d - b}{(a+d)-(b+c)}$$
$$= \frac{1-5}{(2+1)-(5+4)}$$
$$= \frac{-4}{3-9}$$
$$= \frac{-4}{-6}$$
$$= \frac{2}{3}$$
$$1 - r = 1 - \frac{2}{3} = \frac{1}{3}$$

Interpretation

$$p : (1\text{-p}) = \frac{1}{2} : \frac{1}{2}$$

Therefore, out of 2 trials, player X will use his first strategy once and his second strategy once.

$$r : (1\text{-r}) = \frac{2}{3} : \frac{1}{3}$$

Therefore, out of 3 trials, player Y will use his first strategy twice and his second strategy once.

Questions

1. What is a 2x2 game with no saddle point? Explain.
2. Explain the method of solution of a 2x2 game without saddle point.
3. Solve the following game

	Y	
	12	4
X	3	7

Answer: $p = \frac{1}{3}$, $r = \frac{1}{4}$,V=6

4. Solve the following game

	Y	
	5	−4
X	−9	3

Answer: $p = \frac{4}{7}$, $r = \frac{1}{3}$,V=-1

5. Solve the following game

	Y	
X	10	4
	6	8

Answer: $p = \frac{1}{4}$, $r = \frac{1}{2}$,V=7

6. Solve the following game

	Y	
X	20	8
	− 2	10

Answer: $p = \frac{1}{2}$, $r = \frac{1}{2}$, $V = 9$

7. Solve the following game

	Y	
X	10	2
	1	5

Answer: $p = \frac{1}{3}$, $r = \frac{1}{4}$, $V=4$

8. Solve the following game

	Y	
X	12	6
	6	9

Answer: $p = \frac{1}{3}$, $r = \frac{1}{3}$, $V=8$

9. Solve the following game

	Y	
X	10	8
	8	10

Answer: $p = \frac{1}{2}$, $r = \frac{1}{2}$, $V = 9$

10. Solve the following game

	Y	
X	16	4
	4	8

Answer: $p = \frac{1}{4}$, $r = \frac{1}{4}$,V=7

11. Solve the following game

	Y	
X	−11	5
	7	−9

Answer: $p = \frac{1}{2}$, $r = \frac{7}{16}$,V=-2

12. Solve the following game

	Y	
X	−9	3
	5	−7

Answer: $p = \frac{1}{2}$, $r = \frac{5}{12}$,V=-2

****

The Principle Of Dominance

Lesson Outline

- The Principle Of Dominance
- Dividing A Game Into Sub Games

Learning Objectives

After reading this lesson you should be able to

- Understand the principle of dominance
- Solve a game using the principle of dominance
- Solve a game by dividing a game into sub games

The Principle Of Dominance

In the previous lesson, we have discussed the method of solution of a game without a saddle point. While solving a game without a saddle point, one comes across the phenomenon of the dominance of a row over another row or a column over another column in the pay-off matrix of the game. Such a situation is discussed in the sequel.

In a given pay-off matrix A, we say that the ith row dominates the kth row if

$$a_{ij} \geq a_{kj} \quad \text{for all } j = 1,2,\ldots,n$$

and

$$a_{ij} > a_{kj} \quad \text{for at least one } j.$$

In such a situation player A will never use the strategy corresponding to kth row, because he will gain less for choosing such a strategy.

Similarly, we say the pth column in the matrix dominates the qth column if

$$a_{ip} \leq a_{iq} \text{ for all i = 1,2,…,m}$$

and

$$a_{ip} < a_{iq} \text{ for at least one i.}$$

In this case, the player B will loose more by choosing the strategy for the qth column than by choosing the strategy for the pth column. So he will never use the strategy corresponding to the qth column. When dominance of a row (or a column) in the pay-off matrix occurs, we can delete a row (or a column) from that matrix and arrive at a reduced matrix. This principle of dominance can be used in the determination of the solution for a given game.

Let us consider an illustrative example involving the phenomenon of dominance in a game.

Problem 1

Solve the game with the following pay-off matrix

Player B

	I	*II*	*III*	*IV*
Player A	14	2	3	6
	2 3	4	7	5
	6			
	3	3	5	4

Solution

First consider the minimum of each row.

Row	Minimum Value
1	2
2	3
3	3

Maximum of {2, 3, 3} = 3

Next consider the maximum of each column

Column	Maximum Value
1	6
2	4
3	4
4	7
4	6

Minimum of {6, 4, 7, 6}= 4

The following condition holds:

Max {row minima} ≠ min {column maxima}

Therefore we see that there is no saddle point for the game under consideration.

Compare columns II and III

Column II	Column III
2	3
4	7
3	5

We see that each element in column III is greater than the corresponding element in column II. The choice is for player B. Since column II dominates column III, player B will discard his strategy 3.

Now we have the reduced game

	I	*II*	*IV*
1	4	2	6
2	3	4	5
3	6	3	4

For this matrix again, there is no saddle point. Column II dominates column IV. The choice is for player B. So player B will give up his strategy 4

The game reduces to the following:

	I	*II*
1	4	2
2	3	4
3	6	3

This matrix has no saddle point.

The third row dominates the first row. The choice is for player A. He will give up his strategy 1 and retain strategy 3. The game reduces to the following:

$$\begin{matrix} 3 & 4 \\ 6 & 3 \end{matrix}$$

Again, there is no saddle point. We have a 2x2 matrix. Take this matrix as

$$\begin{matrix} a & b \\ c & d \end{matrix}$$

Then we have a = 3, b = 4, c = 6 and d = 3. Use the formulae for p, 1-p, r, 1-r and V.

$$p = \frac{d-c}{(a+d)-(b+c)}$$

$$= \frac{3-6}{(3+3)-(6+4)}$$

$$= \frac{-3}{6-10}$$

$$= \frac{-3}{-4}$$

$$= \frac{3}{4}$$

$$1-p = 1-\frac{3}{4} = \frac{1}{4}$$

$$r = \frac{d-b}{(a+d)-(b+c)}$$

$$= \frac{3-4}{(3+3)-(6+4)}$$

$$= \frac{-1}{6-10}$$

$$= \frac{-1}{-4}$$

$$= \frac{1}{4}$$

$$1-r = 1-\frac{1}{4} = \frac{3}{4}$$

The value of the game

$$V = \frac{ad-bc}{(a+d)-(b+c)}$$

$$= \frac{3 x 3 - 4 x 6}{-4}$$

$$= \frac{-15}{-4}$$

$$= \frac{15}{4}$$

Thus, $X = \left(\frac{3}{4}, \frac{1}{4}, 0, 0\right)$ and $Y = \left(\frac{1}{4}, \frac{3}{4}, 0, 0\right)$ are the optimal strategies.

Method of convex linear combination

A strategy, say s, can also be dominated if it is inferior to a convex linear combination of several other pure strategies. In this case if the domination is strict, then the strategy s can be deleted. If strategy s dominates the convex linear combination of some other pure strategies, then one of the pure strategies involved in the combination may be deleted. The domination will be decided as per the above rules. Let us consider an example to illustrate this case.

Problem 2

Solve the game with the following pay-off matrix for firm A

Firm B

Firm A		B_1	B_2	B_3	B_4	B_5
	A_1	4	8	−2	5	6
	A_2	4	0	6	8	5
	A_3	−2	−6	−4	4	2
	A_4	4	−3	5	6	3
	A_5	4	−1	5	7	3

Solution

First consider the minimum of each row

Row	Minimum Value
1	-2
2	0
3	-6
4	-3
5	-1

Maximum of {-2, 0, -6, -3, -1} = 0

Next consider the maximum of each column.

Column	Maximum Value
1	4
2	8
3	6
4	8
5	6

Minimum of { 4, 8, 6, 8, 6}= 4

Hence,

Maximum of {row minima} ≠ minimum of {column maxima}

So we see that there is no saddle point. Compare the second row with the fifth row. Each element in the second row exceeds the corresponding element in the fifth row. Therefore, A_2 dominates A_5 . The choice is for firm A. It will retain strategy A_2 and give up strategy A_5 . Therefore the game reduces to the following.

$$\begin{array}{c|ccccc} & B_1 & B_2 & B_3 & B_4 & B_5 \\ A_1 & 4 & 8 & -2 & 5 & 6 \\ A_2 & 4 & 0 & 6 & 8 & 5 \\ A_3 & -2 & -6 & -4 & 4 & 2 \\ A_4 & 4 & -3 & 5 & 6 & 3 \end{array}$$

Compare the second and fourth rows. We see that A_2 dominates A_4 . So, firm A will retain the strategy A_2 and give up the strategy A_4 . Thus the game reduces to the following:

	B_1	B_2	B_3	B_4	B_5
A_1	4	8	−2	5	6
A_2	4	0	6	8	5
A_3	−2	−6	−4	4	2

Compare the first and fifth columns. It is observed that B1 dominates B5. The choice is for firm B. It will retain the strategy B_1 and give up the strategy B_5 . Thus the game reduces to the following

	B_1	B_2	B_3	B_4
A_1	4	8	−2	5
A_2	4	0	6	8
A_3	−2	−6	−4	4

Compare the first and fourth columns. We notice that B1 dominates B4. So firm B will discard the strategy B_4 and retain the strategy B_1 . Thus the game reduces to the following:

	B_1	B_2	B_3
A_1	4	8	−2
A_2	4	0	6
A_3	−2	−6	−4

For this reduced game, we check that there is no saddle point.

Now none of the pure strategies of firms A and B is inferior to any of their other strategies. But, we observe that convex linear combination of the strategies B_2 and B_3 dominates B_1 , i.e. the averages of payoffs due to strategies B_2 and B_3 ,

$$\frac{8-2}{2}, \frac{0+6}{2}, \frac{-6-4}{2} = \{3,3,-5\}$$

dominate B_1 . Thus B_1 may be omitted from consideration. So we have the reduced matrix

	B_2	B_3
A_1	8	-2
A_2	0	6
A_3	-6	-4

Here, the average of the pay-offs due to strategies A_1 and A_2 of firm A,

i.e. $\frac{8+0}{2}, \frac{-2+6}{2} = \{4, 2\}$ dominates the pay-off due to A_3. So we get a new reduced 2x2 pay-off matrix.

Firm B's strategy

Firm A's strategy		B_2	B_3
	A_1	8	-2
	A_2	0	6

We have a = 8, b = -2, c = 0 and d = 6

$$p = \frac{d-c}{(a+d)-(b+c)}$$

$$= \frac{6-0}{(6+8)-(-2+0)}$$

$$= \frac{6}{16}$$

$$1 - p = 1 - \frac{3}{8} = \frac{5}{8}$$

$$r = \frac{d-b}{(a+d)-(b+c)}$$

$$= \frac{6-(-2)}{16}$$

$$= \frac{8}{16}$$

$$= \frac{1}{2}$$

$$1 - r = 1 - \frac{1}{2} = \frac{1}{2}$$

Value of the game

$$V = \frac{ad - bc}{(a + d) - (b + c)}$$

$$= \frac{6 x 8 - 0 x(-2)}{16}$$

$$= \frac{48}{16} = 3$$

So the optimal strategies are

$$A = \frac{3}{8}, \frac{5}{8}, 0,0,0 \text{ and } B = 0, \frac{1}{2}, \frac{1}{2}, 0,0 .$$

The value of the game = 3. Thus the game is favourable to firm A.

Problem 3

For the game with the following pay-off matrix, determine the saddle point

Player B

		I	*II*	*III*	*IV*
Player A	1	2	−1	0	−3
	2	1	0	3	2
	3	−3	−2	−1	4

Solution

	Column II	*Column III*	
1	−1	0	0>−1
2	0	3	3 > 0
3	−2	−1	−1> −2

The choice is with the player B. He has to choose between strategies II and III. He will lose more in strategy III than in strategy II, irrespective of what strategy is followed by A. So he will drop strategy III and retain strategy II. Now the given game reduces to the following game.

	I	*II*	*IV*
1	2	−1	−3
2	1	0	2
3	−3	−2	4

Consider the rows and columns of this matrix.

Row minimum:

I Row	:	-3	
II Row :	0		Maximum of {-3, 0, -3} = 0
III Row	:	-3	

Column maximum:

I Column	:	2	
II Column	:	0	Minimum of {2, 0, 4} = 0
III Column	:	4	

We see that

Maximum of row minimum = Minimum of column maximum = 0.
So, a saddle point exists for the given game and the value of the game is 0.

Interpretation

No player gains and no player loses. i.e., The game is not favourable to any player. i.e. It is a fair game.

Problem 4

Solve the game

Player B

$$\text{Player A} \quad \begin{matrix} 4 & 8 & 6 \\ 6 & 2 & 10 \\ 4 & 5 & 7 \end{matrix}$$

Solution

First consider the minimum of each row

Row	Minimum
1	4
2	2
3	4

Maximum of {4, 2, 4} = 4

Next, consider the maximum of each column.

Column	Maximum
1	6
2	8
3	10

Minimum of {6, 8, 10} = 6

Since Maximum of { Row Minima} and Minimum of { Column Maxima } are different, it ~~follows that the given game~~ has no saddle point.

Denote the strategies of player A by A_1 , A_2 , A_3 . Denote the strategies of player B by B_1 , B_2 , B_3 . Compare the first and third columns of the given matrix.

B_1	B_3
4	6
6	10
7	7

The pay-offs in B_3 are greater than or equal to the corresponding pay-offs in B_1 . The player B has to make a choice between his strategies 1 and 3. He will lose more if he follows strategy 3 rather than strategy 1. Therefore he will give up strategy 3 and retain strategy 1. Consequently, the given game is transformed into the following game

	B_1	B_2
A_1	4	8
A_2	6	2
A_3	4	5

Compare the first and third rows of the above matrix

	B_1	B_2
A_1	4	8
A_3	4	5

The pay-offs in A_1 are greater than or equal to the corresponding pay-offs in A_3 . The player A has to make a choice between his strategies 1 and 3. He will gain more if he follows strategy 1 rather than strategy 3. Therefore he will retain strategy 1 and give up strategy 3. Now the given game is transformed into the following game.

	B_1	B_2
A_1	4	8
A_2	6	2

It is a 2x2 game. Consider the row minima

Row	Minimum
1	4
2	2

Maximum of {4, 2} = 4

Next, consider the maximum of each column

Column	Maximum
1	6
2	8

Minimum of {6, 8} = 6

Maximum {row minima} and Minimum {column maxima } are not equal

Therefore, the reduced game has no saddle point. So, it is a mixed game

Take $\begin{matrix} a & b \\ c & d \end{matrix} = \begin{matrix} 4 & 8 \\ 6 & 2 \end{matrix}$. We have a = 4, b = 8, c = 6 and d = 2.

The probability that player A will use his first strategy is p. This is calculated as

$$p = \frac{d - c}{(a + d) - (b + c)}$$

$$= \frac{2-6}{(4+2)-(8+6)}$$

$$= \frac{-4}{6-14}$$

$$= \frac{-4}{-8} = \frac{1}{2}$$

The probability that player B will use his first strategy is r. This is calculated as

$$r = \frac{d - b}{(a + d) - (b + c)}$$

$$= \frac{2 - 8}{-8}$$

$$= \frac{-6}{-8}$$

$$= \frac{3}{4}$$

Value of the game is V. This is calculated as

$$V = \frac{ad - bc}{(a + d) - (b + c)}$$

$$= \frac{4 x 2 - 8 x6}{-8}$$

$$= \frac{8-48}{-8}$$

$$= \frac{-40}{-8} = 5$$

Interpretation

Out of 3 trials, player A will use strategy 1 once and strategy 2 once. Out of 4 trials, player B will use strategy 1 thrice and strategy 2 once. The game is favourable to player A.

Problem 5

Dividing a game into sub-games
Solve the game with the following pay-off matrix.

Player B

Player A		1	2	3
	I	−4	6	3
	II	−3	3	*
	III	2	−3	4

Solution

First, consider the row mimima

Row	Minimum
1	-4
2	-3
3	-3

Maximum of {-4, -3, -3} = -3

Next, consider the column maxima.

Column	Maximum
1	2
2	6
3	4

Minimum of {2, 6, 4} = 2

We see that Maximum of { row minima} ≠ Minimum of { column maxima}.

So the game has no saddle point. Hence it is a mixed game. Compare the first and third columns

$$\begin{array}{ccc} \textit{I Column} & \textit{III Column} & \\ \begin{bmatrix} -4 \\ -3 \\ 2 \end{bmatrix} & \begin{bmatrix} 3 \\ 4 \\ 4 \end{bmatrix} & \begin{array}{r} -4 \le 3 \\ -3 \le 4 \\ 2 \le 4 \end{array} \end{array}$$

We assert that Player B will retain the first strategy and give up the third strategy. We get the following reduced matrix.

$$\begin{array}{ccc} -4 & & 6 \\ -3 & 3 & 2 \\ -3 & & \end{array}$$

We check that it is a game with no saddle point.

Sub games

Let us consider the 2x2 sub games. They are:

-4 $\quad$ $6 -4 6 -3 3$

3

-3 $\quad$ 2 $\quad$ -3 2 -3

First, take the sub game

$-4 \quad 6$

$-3 \quad 3$

Compare the first and second columns. We see that $-4 \le 6$ and

-4

$-3 \le 3$. Therefore, the game reduces to . Since $-4 < -3$, it further

reduces to –3. $\quad -3$

Next, consider the sub game

$-4 \quad 6$

$2 \quad -3$

We see that it is a game with no saddle point. Take a = -4, b = 6, c = 2, d = -3. Then the value of the game is

$-4 \quad 6$

$2 \quad -3$

$-3 \quad 3$

Next, take the sub game . In this case we have a = -3, b = 3, c = 2 -3

2 and d = -3. The value of the game is obtained as

$$V = \frac{ad - bc}{(a + d) - (b + c)}$$

$$= \frac{(-3)(-3) - (3)(2)}{(-3-3)-(3+2)}$$

$$= \frac{9 - 6}{-6-5} = -\frac{3}{11}$$

Let us tabulate the results as follows:

Sub game	Value
$\begin{matrix} -4 & 6 \\ -3 & 3 \end{matrix}$	-3
$\begin{matrix} -4 & 6 \\ 2 & -3 \end{matrix}$	0
$\begin{matrix} -3 & 3 \\ 2 & -3 \end{matrix}$	$-\frac{3}{11}$

The value of 0 will be preferred by the player A. For this value, the first and third strategies of A correspond while the first and second strategies of the player B correspond to the value 0 of the game. So it is a fair game.

Questions

1. Explain the principle of dominance in the theory of games.
2. Explain how a game can be solved through sub games.
3. Solve the following game by the principle of dominance:

Player B

Player A Strategies	I	II	III	IV
1	8	10	9	14
2	10	11	8	12
3	13	12	14	13

Answer: V = 12

4. Solve the game by the principle of dominance:

$$\begin{matrix} 1 & 7 & 2 \\ 6 & 2 & 7 \end{matrix}$$

Answer: V = 4

5. Solve the game with the following pay-off matrix

$$\begin{matrix} 1 & 7 & 2 \\ 6 & 2 & 7 \end{matrix}$$

Answer : $p = \frac{3}{5}, r = \frac{2}{5}, V = -\frac{11}{5}$

6. Solve the game

$$\begin{matrix} 8 & 7 & 6 & -1 & 2 \\ 12 & 10 & 12 & 0 & 4 \\ 14 & 6 & 8 & 14 & 16 \end{matrix}$$

Answer : $p = \frac{4}{9}, r = \frac{7}{9}, V = \frac{70}{9}$

****

Graphical Solution Of A 2X2 Game With No Saddle Point

Lesson Outline

- The principle of graphical solution
- Numerical example

Learning Objectives

After reading this lesson you should be able to

- Understand the principle of graphical solution
- Derive the equations involving probability and expected value
- Solve numerical problems

Example:

Consider the game with the following pay-off matrix.

	Player B	
Player A	2	5
	4	1

First consider the row minima.

Row	Minimum
1	2
2	1

Maximum of {2, 1} = 2.

Next, consider the column maxima.

Column	Maximum
1	4
2	5

Minimum of {4, 5} = 4.

We see that Maximum { row minima} $\neq$ Minimum { column maxima }

So, the game has no saddle point. It is a mixed game.

Equations involving probability and expected value

Let p be the probability that player A will use his first strategy.

Then the probability that A will use his second strategy is 1-p.

Let E be the expected value of pay-off to player A.

When B uses his first strategy

The expected value of pay-off to player A is given by

$$\begin{aligned} E &= 2p + 4(1-p) \\ &= 2p + 4 - 4p \\ &= 4 - 2p \end{aligned} \qquad (1)$$

When B uses his second strategy

The expected value of pay-off to player A is given by

$$\begin{aligned} E &= 5p + 1(1-p) \\ &= 5p + 1 - p \\ &= 4p + 1 \end{aligned} \qquad (2)$$

Consider equations (1) and (2). For plotting the two equations on a graph sheet, get some points on them as follows:

E = -2p+4

p	0	1	0.5
E	4	2	3

E = 4p+1

p	0	1	0.5
E	1	5	3

Graphical solution

Procedure

Take probability and expected value along two rectangular axes in a graph sheet. Draw two straight lines given by the two equations (1) and (2). Determine the point of intersection of the two straight lines in the graph. This will give the common solution of the two equations (1) and (2). Thus we would obtain the value of the game.

Represent the two equations by the two straight lines AB and CD on the graph sheet. Take the point of intersection of AB and CD as T. For this point, we have p = 0.5 and E = 3. Therefore, the value V of the game is 3.

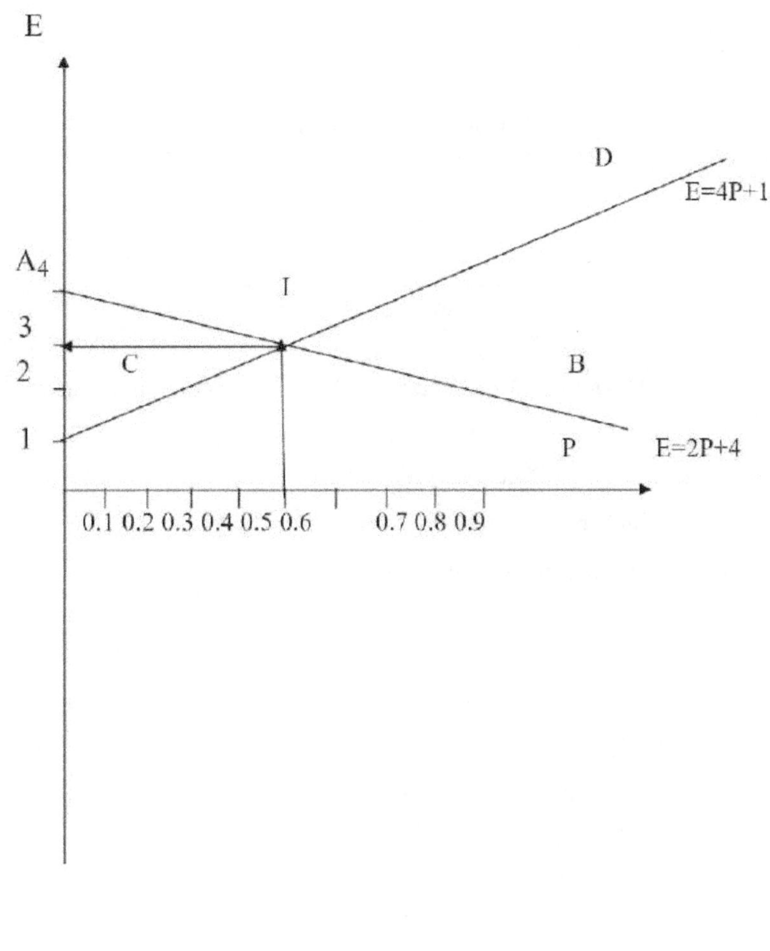

Problem 1

Solve the following game by graphical method.

Player B

Player A	−18	2
	6	−4

Solution

First consider the row minima

Row	Minimum
1	-18
2	-4

Maximum of {-18, - 4} = - 4.

Next, consider the column maxima

Column	Maximum
1	6
2	2

Minimum of {6, 2} = 2

We see that Maximum { row minima} ≠ Minimum { column maxima }

So, the game has no saddle point. It is a mixed game.

Let p be the probability that player A will use his first strategy.

Then the probability that A will use his second strategy is 1-p.

When B uses his first strategy

The expected value of pay-off to player A is given by

$$\begin{aligned} E &= -18\,p + 6\,(1 - p) \\ &= -18\,p + 6 - 6\,p \\ &= -24\,p + 6 \end{aligned} \quad (1)$$

When B uses his second strategy

The expected value of pay-off to player A is given by

$$\begin{aligned} E &= 2\,p - 4\,(1 - p) \\ &= 2\,p - 4 + 4\,p \\ &= 6\,p - 4 \end{aligned} \quad (2)$$

Consider equations (1) and (2). For plotting the two equations on a graph sheet, get some points on them as follows:

E = -24 p + 6

p	0	1	0.5
E	6	-18	-6

E = 6p-4

p	0	1	0.5
E	-4	2	-1

Graphical solution

Take probability and expected value along two rectangular axes in a graph sheet. Draw two straight lines given by the two equations (1) and (2). Determine the point of intersection of the two straight lines in the graph. This will provide the common solution of the two equations (1) and (2). Thus we would get the value of the game.

Represent the two equations by the two straight lines AB and CD on the graph sheet. Take the point of intersection of AB and CD as T. For this point, we have $p = \frac{1}{3}$ and E = -2. Therefore, the value V of the game is -2.

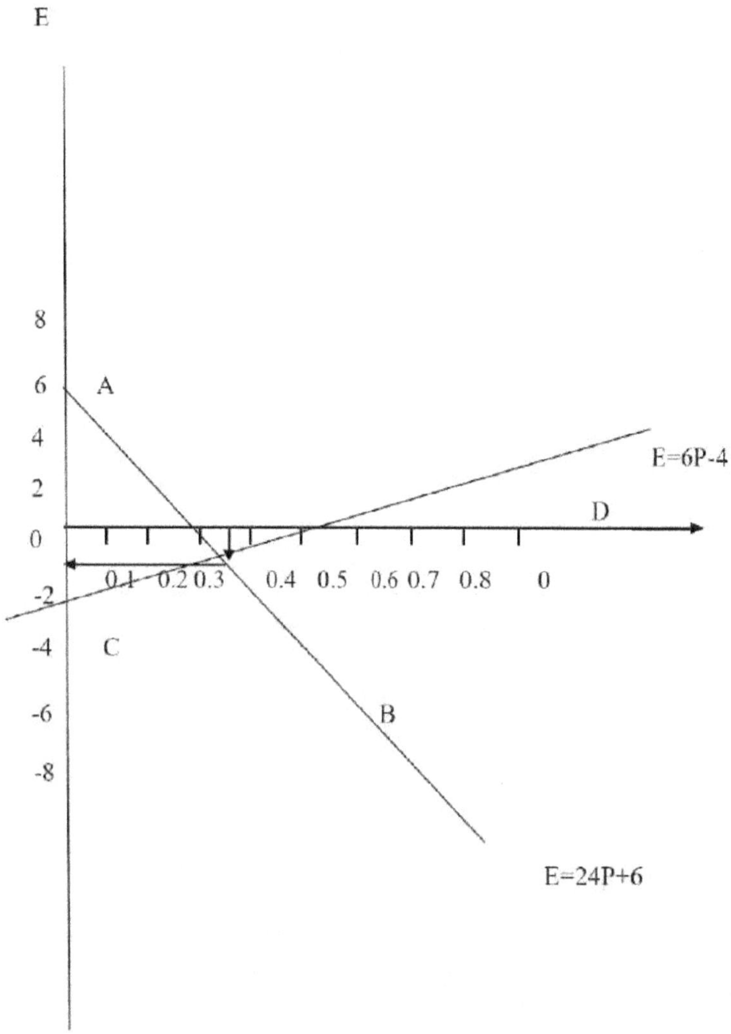

Questions

1. Explain the method of graphical solution of a 2x2 game.

2. Obtain the graphical solution of the game

$$\begin{matrix} 10 & 6 \\ 8 & 12 \end{matrix}$$

Answer: $p = \frac{1}{2}$, $V = 9$

3. Graphically solve the game

$$\begin{matrix} 4 & 10 \\ 8 & 6 \end{matrix}$$

Answer: $p = \frac{1}{4}$, $V = 7$

4. Find the graphical solution of the game

$$\begin{matrix} -12 & 12 \\ 2 & -6 \end{matrix}$$

Answer: $p = \frac{1}{4}$, $V = -\frac{3}{2}$

5. Obtain the graphical solution of the game

$$\begin{matrix} 10 & 6 \\ 8 & 12 \end{matrix}$$

Answer: $p = \frac{1}{2}$, $V = 9$

6. Graphically solve the game

$$\begin{matrix} -3 & -5 \\ -5 & 1 \end{matrix}$$

Answer: $p = \frac{3}{4}$, $V = -\frac{7}{2}$

2 X N Zero-Sum Games

Lesson Outline

- A 2 X N Zero-Sum Game
- Method Of Solution
- Sub Game Approach And Graphical Method
- Numerical Example

Learning Objectives

After reading this lesson you should be able to

- Understand the concept of a 2 x n zero-sum game
- Solve numerical problems

The concept of a 2 x n zero-sum game

When the first player A has exactly two strategies and the second player B has n (where n is three or more) strategies, there results a 2 x n game. It is also called a rectangular game. Since A has two strategies only, he cannot try to give up any one of them. However, since B has many strategies, he can make out some choice among them. He can retain some of the advantageous strategies and discard some disadvantageous strategies. The intention of B is to give as minimum payoff to A as possible. In other words, B will always try to minimize the loss to himself. Therefore, if some strategies are available to B by which he can minimize the payoff to A, then B will retain such strategies and give such strategies by which the payoff will be very high to A.

Approaches for 2 x n zero-sum game

There are two approaches for such games: (1) Sub game approach and (2) Graphical approach.

Sub game approach

The given 2 x n game is divided into 2 x 2 sub games. For this purpose, consider all possible 2 x 2 sub matrices of the payoff matrix of the given game. Solve each sub game and have a list of the values of each sub game. Since B can make out a choice of his strategies, he will discard such of those sub games which result in more payoff to A. On the basis of this consideration, in the long run, he will retain two strategies only and give up the other strategies.

Problem

Solve the following game

Player B

Player A 8-2-69

3

5102

Solution

Let us consider all possible 2x2 sub games of the given game. We have the following sub games:

1.

8 −2

3

5

2.

8 −6

3

10

3.

8 9

3 2

4.

−2	−6
5	10

5.

−2	9
5	2

6.

−6	9
10	2

Let E be the expected value of the pay off to player A. Let p be the probability that player A will use his first strategy. Then the probability that he will use his second strategy is 1-p. We form the equations for E in all the sub games as follows:

Sub game (1)

Equation 1: $E = 8p + 3(1-p) = 5p + 3$
Equation 2: $E = -2p + 5(1-p) = -7p + 5$

Sub game (2)

Equation 1: $E = 8p + 3(1-p) = 5p + 3$
Equation 2: $E = -6p + 10(1-p) = -16p + 10$

Sub game (3)

Equation 1: $E = 8p + 3(1-p) = 5p + 3$
Equation 2: $E = 9p + 2(1-p) = 7p + 2$

Sub game (4)

Equation 1: $E = -2p + 5(1-p) = -7p + 5$
Equation 2: $E = -6p + 10(1-p) = -16p + 10$

Sub game (5)

Equation 1: $E = -2p + 5(1-p) = -7p + 5$
Equation 2: $E = 9p + 2(1-p) = 7p + 2$

Sub game (6)

Equation 1: $E = -6p + 10(1-p) = -16p + 10$
Equation 2: $E = 9p + 2(1-p) = 7p + 2$

Solve the equations for each sub game. Let us tabulate the results for the various sub games. We have the following:

Sub game	p	Expected value E
1	1/6	23/6
2	1/3	14/3
3	1/2	11/2
4	5/9	10/9
5	3/14	7/2
6	8/23	102/23

Interpretation

Since player A has only 2 strategies, he cannot make any choice on the strategies. On the other hand, player B has 4 strategies. Therefore he can retain any 2 strategies and give up the other 2 strategies. This he will do in such a way that the pay-off to player A is at the minimum. The pay-off to A is the minimum in the case of sub game 4. i.e., the sub game with the matrix

.

$$\begin{matrix} -2 & -6 \\ 5 & 10 \end{matrix}$$

Therefore, in the long run, player B will retain his strategies 2 and 3 and give up his strategies 1 and 4. In that case, the probability that A will use his first strategy is p = 5/9 and the probability that he will use his second strategy is 1-p = 4/9 . i.e., Out of a total of 9 trials, he will use his first strategy five times and the second strategy four times. The value of the game is10/9 . The positive sign of V shows that the game is favourable to player A.

Graphical Solution

Now we consider the graphical method of solution to the given game.

Draw two vertical lines MN and RS. Note that they are parallel to each other. Draw UV perpendicular to MN as well as RS. Take U as the origin on the line MN. Take V as the origin on the line RS.

Mark units on MN and RS with equal scale. The units on the two lines MN and RS are taken as the payoff numbers. The payoffs in the first row of the given matrix are taken along the line MN while the payoffs in the second row are taken along the line RS.

We have to plot the following points: (8, 3), (-2, 5), (-6, 10), (9, 2). The points 8, -2, -6, 9 are marked on MN. The points 3, 5, 10, 2 are marked on RS.

Join a point on MN with the corresponding point on RS by a straight line. For example, join the point 8 on MN with the point 3 on RS. We have 4 such straight lines. They represent the 4 moves of the second player. They intersect in 6 points. Take the lowermost point of intersection of the straight lines. It is called the Maximin point. With the help of this point, identify the optimal strategies for the second player. This point corresponds to the points −2 and −6 on MN and 5 and 10 on RS. They correspond to the sub game with the matrix .

$$\begin{matrix} -2 & -6 \\ 5 & 10 \end{matrix}$$

The points −2 and −6 on MN correspond to the second and third strategies of the second player. Therefore, the graphical method implies that, in the long run, the second player will retain his strategies 2 and 3 and give up his strategies 1 and 4.

We graphically solve the sub game with the above matrix. We have to solve the two equations $E = -7p + 5$ and $E = -16p + 10$. Represent the two equations by two straight lines AB and CD on the graph sheet. Take the point of intersection of AB and CD as T. For this point, we have $p = 5/9$ and $E = 10/9$. Therefore, the value V of the game is 10/9 . We see that the probability that first player will use his first strategy is p = 8/9 and the probability that he will use his second strategy is 1-p = 4/9 .

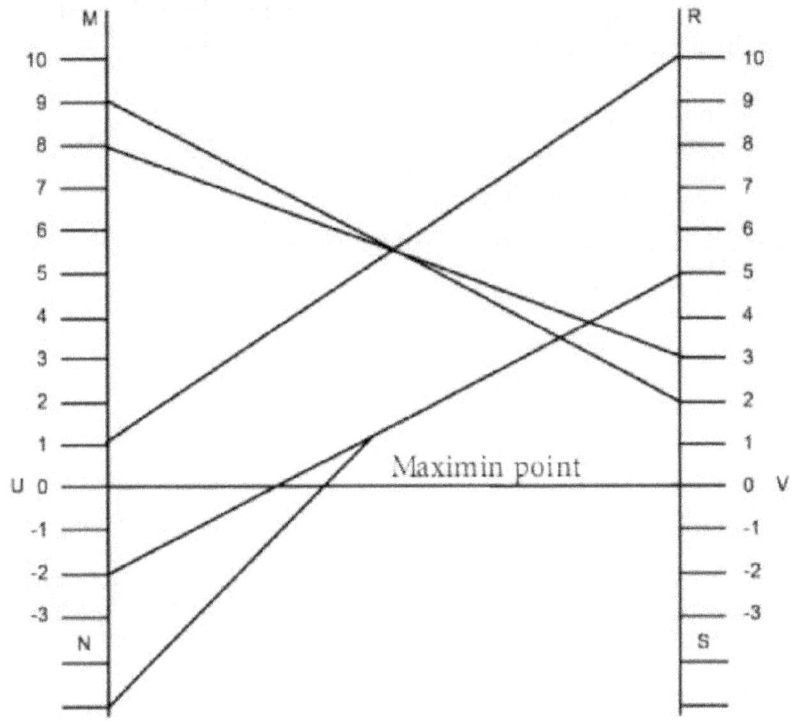

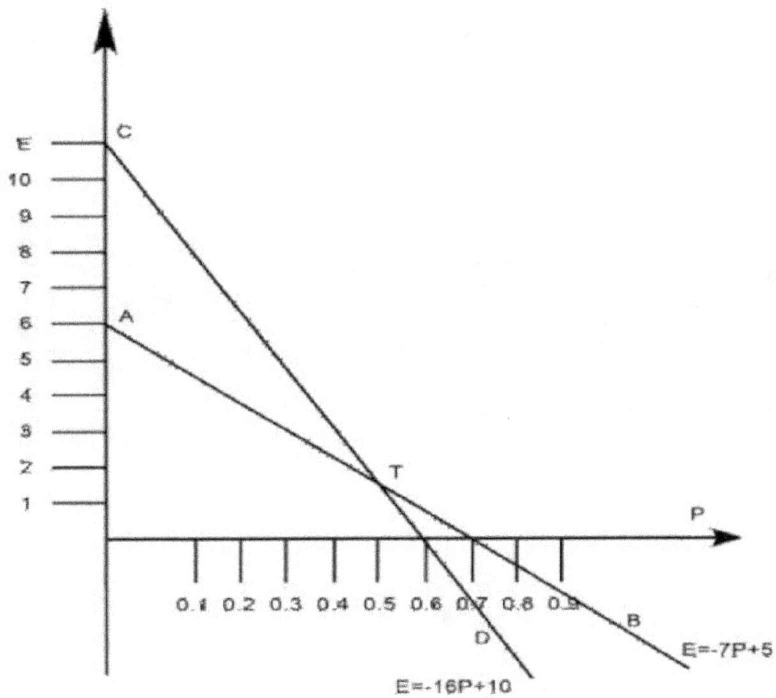

$E = -7p+5$

p	0	1	0.5
E	5	- 2	1.5

$E = -16p+10$

p	0	1	0.5
E	10	- 6	2

Questions

1. Explain a 2 x n zero-sum game.
2. Describe the method of solution of a 2 x n zero-sum game.
3. Solve the following game:

	Player B		
Player A	10	2	6
	1	5	8

Answer: p = , V = 4t

****

Lesson 7 - M X 2 Zero-Sum Games

Lesson Outline

- An m x 2 zero-sum game
- Method of solution
- Sub game approach and graphical method
- Numerical example

Learning Objectives

After reading this lesson you should be able to

- Understand the concept of an m x 2 zero-sum game
- Solve numerical problems

The concept of an m x 2 zero-sum game

When the second player B has exactly two strategies and the first player A has m (where m is three or more) strategies, there results an m x 2 game. It is also called a rectangular game. Since B has two strategies only, he will find it difficult to discard any one of them. However, since A has more strategies, he will be in a position to make out some choice among them. He can retain some of the most advantageous strategies and give up some other strategies. The motive of A is to get as maximum payoff as possible. Therefore, if some strategies are available to A by which he can get more payoff to himself, then he will retain such strategies and discard some other strategies which result in relatively less payoff.

Approaches for m x 2 zero-sum game

There are two approaches for such games: (1) Sub game approach and (2) Graphical approach.

Sub game approach

The given m x 2 game is divided into 2 x 2 sub games. For this purpose, consider all possible 2 x 2 sub matrices of the payoff matrix of the given game. Solve each sub game and have a list of the values of each sub game. Since A can make out a choice of his strategies, he will be interested in such of those sub games which result in more payoff to himself. On the basis of this consideration, in the long run, he will retain two strategies only and give up the other strategies.

Problem

Solve the following game

Player B
Strategies

		I	*II*
Player A Strategies	1	5	8
	2	−210	
	3	12	4
	4	6	5

Solution

Let us consider all possible 2x2 sub games of the given game. We have the following sub games:

7. $\begin{bmatrix} 5 & 8 \\ -2 & 10 \end{bmatrix}$ 8. $\begin{bmatrix} 5 & 8 \\ 12 & 4 \end{bmatrix}$

9. $\begin{bmatrix} 5 & 8 \\ 6 & 5 \end{bmatrix}$ 10. $\begin{bmatrix} -2 & 10 \\ 6 & 5 \end{bmatrix}$

11. $\begin{bmatrix} -2 & 10 \\ 6 & 5 \end{bmatrix}$ 12. $\begin{bmatrix} 12 & 4 \\ 6 & 5 \end{bmatrix}$

Let E be the expected value of the payoff to player A. i.e., the loss to player I. Let r be the probability that player B will use his first strategy. Then the probability that he will use his second strategy is 1-r. We form the equations for E in all the sub games as follows:

Sub game (1)

Equation 1: $E = 5r + 8(1 - r) = -3r + 8$

Equation 2: $E = -2r + 10(1 - r) = -12r + 10$

Sub game (2)

Equation 1: $E = 5r + 8(1 - r) = -3r + 8$

Equation 2: $E = 12r + 4(1 - r) = 8r + 4$

Sub game (3)

Equation 1: $E = 5r + 8(1 - r) = -3r + 8$

Equation 2: $E = 6r + 5(1 - r) = r + 5$

Sub game (4)

Equation 1: $E = -2r + 10(1 - r) = -12r + 10$

Equation 2: $E = 12r + 4(1 - r) = 8r + 4$

Sub game (5)

Equation 1: $E = -2r + 10(1 - r) = -12r + 10$

Equation 2: $E = 6r + 5(1 - r) = r + 5$

Sub game (6)

Equation 1: $E = 12r + 4(1 - r) = 8r + 4$

Equation 2: $E = 6r + 5(1 - r) = r + 5$

Solve the equations for each 2x2 sub game. Let us tabulate the results for the various sub games. We have the following: Sub game

Sub game	R	Expected value E
1	2/9	22/9
2	4/11	76/11
3	3/4	23/4
4	3/10	32/5
5	5/13	70/13
6	1/7	36/7

Interpretation

Since player B has only 2 strategies, he cannot make any choice on his strategies. On the other hand, player A has 4 strategies and so he can retain any 2 strategies and give up the other 2 strategies. Since the choice is with A, he will try to maximize the payoff to himself. The pay-off to A is the maximum in the case of sub game 1. i.e., the sub game with the matrix .

$$\begin{matrix} 5 & 8 \\ -2 & 10 \end{matrix}$$

Therefore, player A will retain his strategies 1 and 2 and discard his strategies 3 and 4, in the long run. In that case, the probability that B will use his first strategy is r = 2/9 and the probability that he will use his second strategy is 1-r = 7/9 . i.e., Out of a total of 9 trials, he will use his first strategy two times and the second strategy seven times.

The value of the game is 22/3 . The positive sign of V shows that the game is favourable to player A.

Graphical Solution

Now we consider the graphical method of solution to the given game.

Draw two vertical lines MN and RS. Note that they are parallel to each other. Draw UV perpendicular to MN as well as RS. Take U as the origin on the line MN. Take V as the origin on the line RS.

Mark units on MN and RS with equal scale. The units on the two lines MN and RS are taken as the payoff numbers. The payoffs in the first row of the given matrix are taken along the line MN while the payoffs in the second row are taken along the line RS.

We have to plot the following points: (5, 8), (-2, 10), (12, 4), (6, 5).The points 5, -2, 12, 6 are marked on MN. The points 8, 10, 4, 5 are marked on RS.

Join a point on MN with the corresponding point on RS by a straight line. For example, join the point 5 on MN with the point 8 on RS. We have 4 such straight lines. They represent the 4 moves of the first player. They intersect in 6 points. Take the uppermost point of intersection of the straight lines. It is called the Minimax point. With the help of this point, identify the optimal strategies for the first player. This point corresponds to the points 5 and -2 on MN and 8 and 10 on RS. They correspond to the sub game with the matrix

$$\begin{matrix} 5 & 8 \\ -2 & 10 \end{matrix}$$

. The points 5 and -2 on MN correspond to the first and second strategies of the first player. Therefore, the graphical method implies that the first player will retain his strategies 1 and 2 and give up his strategies 3 and 4, in the long run.

We graphically solve the sub game with the above matrix. We have to solve the two equations E = - 3 r + 8 and E = - 12 r + 10. Represent the two equations by two straight lines AB and CD on the graph sheet. Take the point of intersection of AB and CD as T. For this point, we have r = 2/9 and E = 22/3. Therefore, the value V of the game is 22/3 . We see that the probability that the second player will use his first strategy is r = 2/9 and the probability that he will use his second strategy is 1-r = 7/9 .

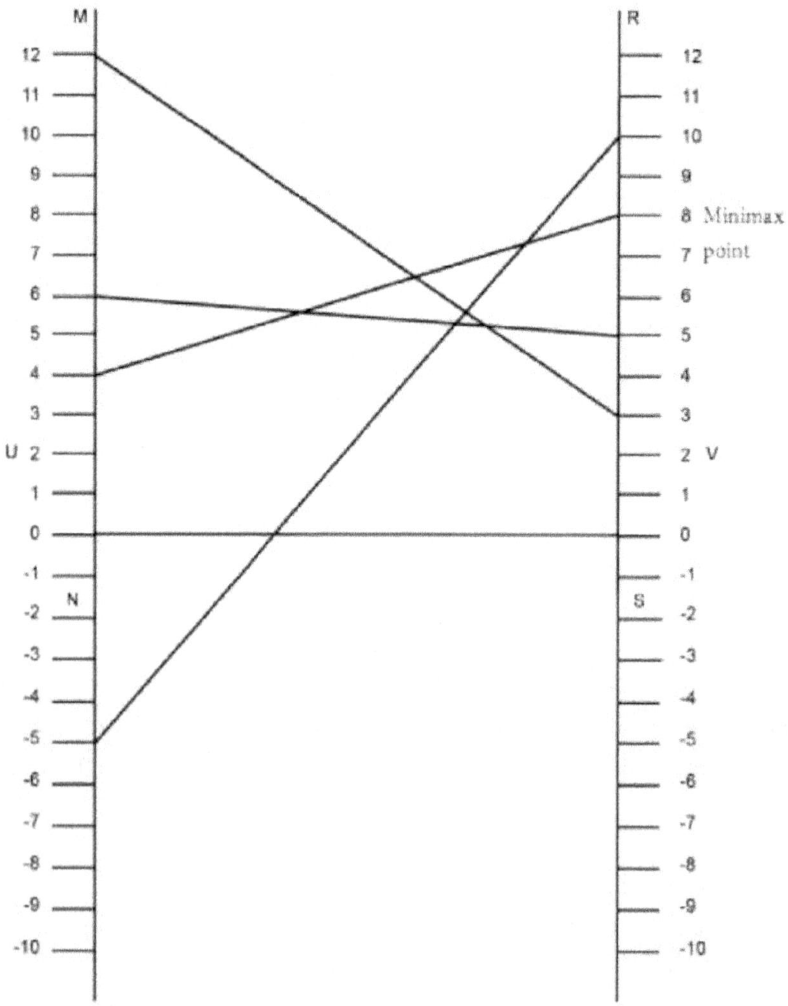
M
R
12
11
10
9
8
7
6
5
4
3
U 2
1
0
-1
N
-2
-3
-4
-5
-6
-7
-8
-9
-10
S
2 V
8 Minimax
7 point

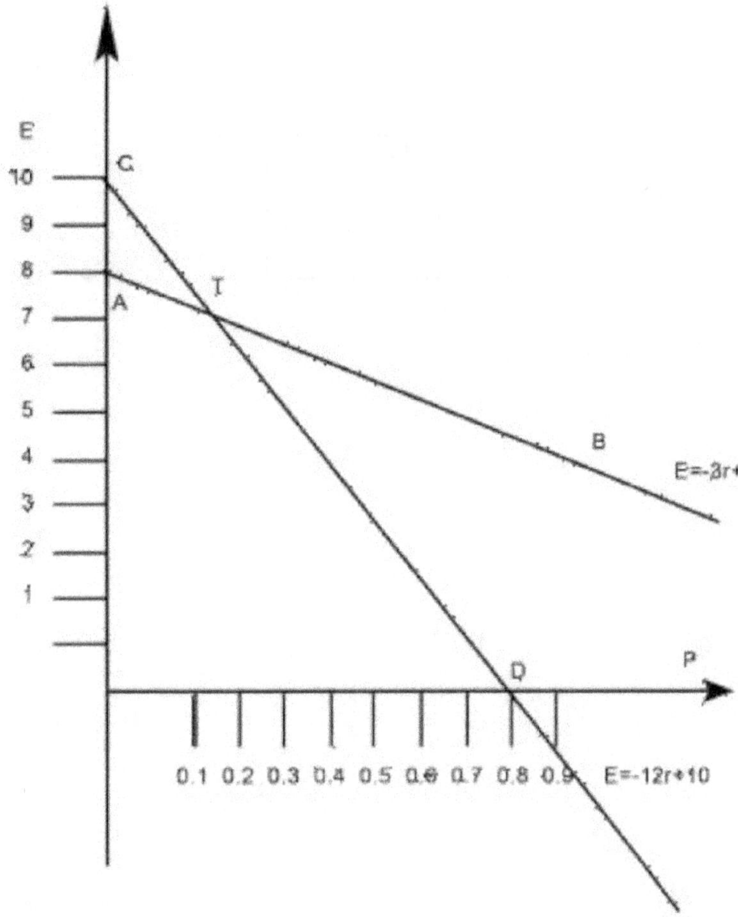

E = - 3r+8

p	0	1	0.5
E	8	5	6.5

E = - 12r+10

p	0	1	0.5
E	10	- 2	4

Questions

3. What is an m x 2 zero-sum game? Explain.
4. How will you solve an m x 2 zero-sum game? Explain.
5. Solve the following game:

Player B

Strategies

		I	*II*
Player A Strategies	1	20	8
	2	5	2
	3	8	12

Answer: r = , V = 11

****

Linear Programming Approach To Game Theory

Lesson Outline

How to solve a game with LPP?

Formulation of LPP

Solution by simplex method

Learning Objectives

After reading this lesson you should be able to

Understand the transformation of a game into LPP

Solve a game by simplex method

Introduction

When there is neither saddle point nor dominance in a problem of game theory and the payoff matrix is of order 3x3 or higher, the probability and graphical methods cannot be employed. In such a case, linear programming approach may be followed to solve the game.

Linear programming technique

A general approach to solve a game by linear programming technique is presented below. Consider the following game:

Player B

		Y_1	Y_2	Y_n
Player A	X_1	a_{11}	a_{12}	a_{1n}
	X_2	a_{21}	a_{22}	a_{2n}
	X_m	a_{m1}	a_{m2}	a_{mn}

It is required to determine the optimal strategy for A = $\{X_1, X_2, \ldots X_m\}$ and B = $\{Y_1, Y_2, \ldots Y_n\}$. First we shall determine the optimal strategies of player B.

If player A adopts strategy , then the expected value of loss to B is ,

$$a_{11}Y_1 + a_{12}Y_2 + \ldots + a_{1n}Y_n \leq V$$

where V is the value of game. If A adopts strategy X_2 , then the expected value of loss to B is

$$a_{21}Y_1 + a_{22}Y_2 + \ldots + a_{2n}Y_n \leq V$$

and so on. Also we have

$$Y_1 + Y_2 + \ldots + Y_n = 1$$

and

$Y_j \geq 0$ for all j.

Without loss of generality, we can assume that V > 0. Divide each of the above relation by V and let $Y'_j = \frac{Y_j}{V}$.

Then we have

$$\sum Y'_j = \sum \frac{Y_j}{V} = \frac{1}{V}$$

From this we obtain

$$a_{11}Y'_1 + a_{12}Y'_2 + \ldots + a_{1n}Y'_n \leq 1,$$
$$a_{21}Y'_1 + a_{22}Y'_2 + \ldots + a_{2n}Y'_n \leq 1,$$

$$a_{m1}Y'_1 + a_{m2}Y'_2 + \ldots + a_{nm}Y'_n \leq 1$$

and

$$Y'_1 + Y'_2 + \dots + Y'_n = \frac{1}{V}$$

with $Y'_j \geq 0$ for all j.

The objective of player B is to minimise the loss to himself. Thus the problem is to minimize V, or equivalently to maximise $\frac{1}{V}$. Therefore, the objective of player B is to maximise the value of $Y'_1 + Y'_2 + \dots + Y'_n$ subject to the m linear constraints provided above.

Statement of the problem

Maximise: $Y'_1 + Y'_2 + \dots + Y'_n$, subject to

$$a_{11}Y'_1 + a_{12}Y'_2 + \dots + a_{1n}Y'_n \leq 1,$$
$$a_{21}Y'_1 + a_{22}Y'_2 + \dots + a_{2n}Y'_n \leq 1,$$

$$a_{m1}Y'_1 + a_{m2}Y'_2 + \dots + a_{mn}Y'_n \leq 1$$

$$Y'_1, Y'_2, \dots, Y'_n \geq 0.$$

We can use simplex method to solve the above problem. For this purpose, we have to introduce non-negative slack variables $s_1, s_2, \dots, s_m$ to each of the inequalities. So the problem can be restated as follows:
Restatement of the problem:

Maximise: $Y'_1 + Y'_2 + \dots + Y'_n + 0\,s_1 + 0\,s_2 + \dots + 0s_m$ subject to

$$a_{11}Y'_1 + a_{12}Y'_2 + \dots + a_{1n}Y'_n + s_1 = 1,$$
$$a_{21}Y'_1 + a_{22}Y'_2 + \dots + a_{2n}Y'_n + s_2 = 1,$$

$$a_{m1}Y'_1 + a_{m2}Y'_2 + \dots + a_{mn}Y'_n + s_m = 1$$

with $Y_j = Y'_j\ V$ for all j and $s_1 \geq 0, s_2 \geq 0, \ldots, s_m \geq 0$.

Thus we get the optimal strategy for player B to be (Y_1 , Y_2 ,...,Y_n).

In a similar manner we can determine the optimal strategy for player A.

Application

We illustrate the method for a 2X2 zero sum game.

Problem 1

Solve the following game by simplex method for LPP:

Player B

Player A	
	6
	2

Solution

Row minima :	I row	: 3
	II row	: 2
	Maximum of {3,2} = 3	
Column maxima:	I column	: 5
	II column	: 6
	Minimum of {5,6} = 5	

So, Maximum of {Row minima} $\neq$ Minimum of {Column maxima}.

Therefore the given game has no saddle point. It is a mixed game. Let us convert the given game into a LPP.

Problem formulation

Let V denote the value of the game. Let the probability that the player B will use his first strategy be r and second strategy be s. Let V denote the value of the game.

When A follows his first strategy

The expected payoff to A (i.e., the expected loss to B) = 3 r + 6 s.

This pay-off cannot exceed V. So we have =3 r + 6 s V (1)

When A follows his second strategy

The expected pay-off to A (i.e., expected loss to B) = 5 r + 2 s.

This cannot exceed V. Hence we obtain the condition = 5 r + 2 s V (2)

From (1) and (2) we have

$$\frac{r}{V} + 6\frac{s}{V} \le 1$$

$$\text{and } 5\frac{r}{V} + 2\frac{s}{V} \le 1$$

Substitute $\frac{r}{V} = x, \ \frac{s}{V} = y.$

Then we have

$$3x + 6y \le 1$$

$$\text{and } 5x + 2y \le 1$$

where r and s are connected by the relation

$$r + s = 1$$

$$\text{i.e., } \frac{r}{V} + \frac{s}{V} = \frac{1}{V}$$

$$\text{i.e., } x + y = \frac{1}{V}$$

B will try to minimise V. i.e., He will try to maximise $\frac{1}{V}$. Thus we have the following LPP.

Maximize $\frac{1}{V} = x + y,$

subject to the restrictions

$$3x + 6y \le 1,$$

$$5x + 2y \le 1,$$

$$x \ge 0, y \ge 0$$

Solution of LPP

Introduce two slack variables s_1 , s_2 . Then the problem is transformed into the
following one:

Maximize $\frac{1}{V} = x + y + 0.s_1 + 0.s_2$

subject to the constraints

$$3x + 6y + 1.s_1 + 0.s_2 = 1,$$
$$5x + 2y + 0.s_1 + 1.s_2 = 1,$$
$$x \geq 0, y \geq 0, s_1 \geq 0, s_2 \geq 0$$

Let us note that the above equations can be written in the form of a single matrix equation as

AX=B

where $A = \begin{bmatrix} 3 & 6 & 1 & 0 \\ 5 & 2 & 0 & 1 \end{bmatrix}$, $X = \begin{bmatrix} x \\ y \\ s_1 \\ s_2 \end{bmatrix}$, $B = \begin{bmatrix} 1 \\ 1 \end{bmatrix}$

The entries in B are referred to as the b – values. Initially, the basic variables are s_1 , s_2 . We have the following simplex tableau:

	x	y	s_1	s_2	b – value
s_1 - row	3	6	1	0	1
s_2 - row	5	2	0	1	1
Objective function row	-1	-1	0	0	0

Consider the negative elements in the objective function row. They are –1, -1. The absolute values are 1, 1. There is a tie between these coefficients. To resolve the tie, we select the variable x. We take the new basic variable as x. Consider the ratio of b-value to x-value. We have the following ratios:

s_1 - row $\quad : \frac{1}{3}$

s_2 - row : $\frac{1}{5}$

Minimum of { $\frac{1}{3}, \frac{1}{5}$ } = $\frac{1}{5}$.

Hence select as the leaving variable. Thus the pivotal element is 5. We obtain the following tableau at the end of Iteration No. 1.

	x	y	s_1	s_2	b – value
s_1 - row	0	$\frac{24}{5}$	1	$-\frac{3}{5}$	$\frac{2}{5}$
x - row	1	$\frac{2}{5}$	0	$\frac{1}{5}$	$\frac{1}{5}$
Objective function row	0	$-\frac{3}{5}$	0	$\frac{1}{5}$	$\frac{1}{5}$

Now, the negative element in the objective function row is $-\frac{3}{5}$. This corresponds to y. We take the new basic variable as y. Consider the ratio of b-value to y-value. We have the following ratios:

s_1 - row : $\frac{2}{5} \Big/ \frac{24}{5} = \frac{1}{12}$

x - row : $\frac{1}{5} \Big/ \frac{2}{5} = \frac{1}{2}$

Minimum of $\frac{1}{12}$ $\frac{1}{2}$ $= \frac{1}{12}$

Hence select s_1 as the leaving variable. The pivotal element is get $\frac{24}{5}$. We the following tableau at the end of Iteration No. 2.

	x	y	s_1	s_2	b – value
y - row	0	1	$\frac{5}{24}$	$-\frac{1}{8}$	$\frac{1}{12}$
x - row	1	0	$-\frac{1}{12}$	$\frac{1}{4}$	$\frac{1}{6}$
Objective function row	0	0	$\frac{1}{8}$	$\frac{1}{8}$	$\frac{1}{4}$

Since both x and y have been made basic variables, we have reached the stopping condition.

The optimum value of $\frac{1}{V}$ is $\frac{1}{4}$. This is provided by x = $\frac{1}{6}$ and y = $\frac{1}{12}$. Thus the optimum value of the game is obtained as V = 4. Using the relations $\frac{r}{V} = x, \frac{s}{V} = y$, we obtain $r = \frac{4}{6} = \frac{2}{3}$ and $s = \frac{4}{12} = \frac{1}{3}$.

Problem 2

Solve the following game

Player B

Player A

Player A	2	5
	4	1

Solution

The game has no saddle point. It is a mixed game. Let the probability that B will use his first strategy be r. Let the probability that B will use his second strategy be s. Let V be the value of the game.

When A follows his first strategy

The expected payoff to A (i.e., the expected loss to B) = 2 r +5 s. The pay-off to A cannot exceed V. So we have = 2 r + 5 s V (I)

When A follows his second strategy

The expected pay-off to A (i.e., expected loss to B) = 4 r + s.
The pay-off to A cannot exceed V. Hence we obtain the condition

$$= 4r + s \; V \; (II)$$

From (I) and (II) we have

$$2\frac{r}{V} + 5\frac{s}{V} \leq 1$$

$$and\; 4\frac{r}{V} + \frac{s}{V} \leq 1$$

Substitute

$$\frac{r}{V} = x \; and \; \frac{s}{V} = y.$$

Thus we have

$$2x + 5y \leq 1$$

$$and\; 4x + y \leq 1$$

where r and s are connected by the relation

$$r + s = 1$$

i.e.,
$$\frac{r}{V} + \frac{s}{V} = \frac{1}{V}$$

i.e.,
$$x + y = \frac{1}{V}$$

The objective of B is to minimise V. i.e., He will try to maximise $\frac{1}{V}$.
Thus we are led to the following linear programming problem:

Maximize $\frac{1}{V} = x + y$

subject to the constraints
$$2x + 5y \leq 1,$$
$$4x + y \leq 1,$$
$$x \geq 0, y \geq 0.$$

To solve this linear programming problem, we use simplex method as detailed below.

Introduce two slack variables s_1, s_2. Then the problem is transformed into the following one:

Maximize
$$\frac{1}{V} = x + y + 0.s_1 + 0.s_2$$

$$2x + 5y + 1.s_1 + 0.s_2 = 1,$$

subject to the constraints $\quad 4x + y + 0.s_1 + 1.s_2 = 1,$

$$x \geq 0, y \geq 0, s_1 \geq 0, s_2 \geq 0$$

We have the following simplex tableau:

	x	y	s_1	s_2	b – value
s_1 - row	2	5	1	0	1
s_2 - row	4	1	0	1	1
Objective function row	-1	-1	0	0	0

Consider the negative elements in the objective function row. They are –1, -1. The absolute value are 1, 1. There is a tie between these coefficients. To resolve the tie, we select the variable x. We take the new basic variable as x. Consider the ratio of b-value to x-value. We have the following ratios:

s_1 - row $\quad : \frac{1}{2}$

s_2 - row $\quad : \frac{1}{4}$

Minimum of $\{ \frac{1}{2}, \frac{1}{4} \} = \frac{1}{4}$

Hence select s_2 as the leaving variable. Thus the pivotal element is 4. We obtain the following tableau at the end of Iteration No. 1.

	x	y	s_1	s_2	b – value
s_1 - row	0	$\frac{9}{2}$	1	$-\frac{1}{2}$	$\frac{1}{2}$
x - row	1	$\frac{1}{4}$	0	$\frac{1}{4}$	$\frac{1}{4}$
Objective function row	0	$-\frac{3}{4}$	0	$\frac{1}{4}$	$\frac{1}{4}$

Now, the negative element in the objective function row is $-\frac{3}{4}$. This corresponds to y. We take the new basic variable as y. Consider the ratio of b-value to y-value. We have the following ratios:

s_1 - row : $\dfrac{\frac{1}{2}}{\frac{9}{2}} = \dfrac{1}{9}$

x - row : $\dfrac{\frac{1}{4}}{\frac{1}{4}} = 1$

Minimum of $\dfrac{1}{9}, 1 = \dfrac{1}{9}$

Hence select s_1 as the leaving variable. The pivotal element is $\frac{9}{2}$. We get the following tableau at the end of Iteration No. 2.

	x	y	s_1	s_2	b – value
y - row	0	1	$\frac{2}{9}$	$-\frac{1}{9}$	$\frac{1}{9}$
x - row	1	0	$-\frac{1}{18}$	$\frac{5}{18}$	$\frac{2}{9}$
Objective function row	0	0	$\frac{1}{6}$	$\frac{1}{6}$	$\frac{1}{3}$

Since both x and y have been made basic variables, we have reached the stopping condition.

The optimum value of $\frac{1}{V}$ is $\frac{1}{3}$.

This is provided by x = $\frac{2}{9}$ and y = $\frac{1}{9}$. Thus the optimum value of the game is got as V = 3. Using the relations $\frac{r}{V} = x$, $\frac{s}{V} = y$, we obtain $r = \frac{6}{9} = \frac{2}{3}$ and $s = \frac{3}{9} = \frac{1}{3}$.

Problem 3

Solve the following game by simplex method for LPP

Player B

Player A		
	−48	2
		−4

Solution

Row minima : I row : -48

II row : -4

Maximum of {-48, -4} = -4

Column maxima: I column : 6

II column : 2

Minimum of (6, 2} = 2

So, Maximum of {Row minima} ≠ Minimum of {Column maxima}.

Therefore the given game has no saddle point. It is a mixed game. Let us convert the given game into a LPP.

Problem formulation

Let V denote the value of the game. Let the probability that the player B will use his first strategy be r and second strategy be s. Let V denote the value of the game.

When A follows his first strategy

The expected payoff to A (i.e., the expected loss to B) = - 48 r + 2 s.
This pay-off cannot exceed V. So we have =- 48 r + 2 s V (1)'

When A follows his second strategy

The expected pay-off to A (i.e., expected loss to B) = 6 r - 4 s. This cannot exceed V. Hence we obtain the condition = 6 r - 4 s V (2)'

From (1)' and (2)' we have

$$-48\frac{r}{V} + 2\frac{s}{V} \leq 1$$

and

$$6\frac{r}{V} - 4\frac{s}{V} \leq 1$$

Substitute $\frac{r}{V} = x$, $\frac{s}{V} = y$.

Thus we have
$$-48x + 2y \le 1$$
$$and \;\; 6x - 4y \le 1$$

where r and s are connected by the relation

$$r + s = 1 .$$

i.e., $$\frac{r}{V} + \frac{s}{V} = \frac{1}{V}$$

i.e., $$x + y = \frac{1}{V}$$

B will try to minimise V. i.e., He will try to maximise $\frac{1}{V}$. Thus we have the following LPP.

Maximize $$\frac{1}{V} = x + y,$$

subject to the restrictions
$$-48x + 2y \le 1,$$
$$6x - 4y \le 1,$$
$$x \ge 0, y \ge 0$$

Solution of LPP

Introduce two slack variables s_1, s_2. Then the problem is transformed into the following one:

Maximize $$\frac{1}{V} = x + y + 0.s_1 + 0.s_2$$

subject to the constraints
$$-48x + 2y + 1.s_1 + 0.s_2 = 1,$$
$$6x - 4y + 0.s_1 + 1.s_2 = 1,$$
$$x \ge 0, y \ge 0, s_1 \ge 0, s_2 \ge 0$$

Initially, the basic variables are s_1, s_2. We have the following simplex tableau:

	x	y	s_1	s_2	b – value
s_1 - row	-48	2	1	0	1
s_2 - row	6	-4	0	1	1
Objective function row	-1	-1	0	0	0

Consider the negative elements in the objective function row. They are –1, -1. The absolute value are 1, 1. There is a tie between these coefficients. To resolve the tie, we select the variable x. We take the new basic variable as x. Consider the ratio of b-value to x-value. We have the following ratios:

s_1 - row : $-\frac{1}{48}$

s_2 - row: $\frac{1}{6}$

Minimum of $\{ -\frac{1}{48}, \frac{1}{6} \} = -\frac{1}{48}$.

Hence select s_1 as the leaving variable. Thus the pivotal element is - 48. We obtain the following tableau at the end of Iteration No. 1.

	x	y	s_1	s_2	b – value
x - row	1	$-\frac{1}{24}$	$-\frac{1}{48}$	0	$-\frac{1}{48}$
s_2 - row	0	$-\frac{15}{4}$	$\frac{1}{8}$	1	$\frac{9}{8}$
Objective function row	0	$-\frac{25}{24}$	$-\frac{1}{48}$	0	$-\frac{1}{48}$

Now, the negative element in the objective function row is $-\frac{25}{24}$. This corresponds to y. We take the new basic variable as y. Consider the ratio of b-value to y-value. We have the following ratios:

- row : $\frac{-\frac{1}{48}}{-\frac{1}{24}} = \frac{1}{2}$

s_2 - row : $\frac{\frac{9}{82}}{-\frac{15}{4}} = -\frac{3}{10}$

Minimum of $\frac{1}{2}, -\frac{3}{10} = -\frac{3}{10}$

Hence select s_2 as the leaving variable. The pivotal element is $-\frac{15}{4}$. We get the following tableau at the end of Iteration No. 2.

	x	y	s_1	s_2	b – value
x - row	1	0	$-\frac{1}{480}$	$-\frac{1}{90}$	$-\frac{1}{30}$
y - row	0	1	$-\frac{1}{30}$	$-\frac{4}{15}$	$-\frac{3}{10}$
Objective function row	0	0	$-\frac{1}{18}$	$-\frac{5}{18}$	$-\frac{1}{3}$

Since both x and y have been made basic variables, we have reached the stopping condition.

The optimum value of $\frac{1}{V}$ is $-\frac{1}{3}$. This is provided by x = $-\frac{1}{30}$ and y = $-\frac{3}{10}$.

Thus the optimum value of the game is got as V = - 3. Using the relations $\frac{r}{V} = x$, $\frac{s}{V} = y$, we obtain $r = \frac{1}{10}$ and $s = \frac{9}{10}$.

Problem 4

Transform the following game into an LPP

$$= \frac{9}{10}$$

Solution

We have to determine the optimal strategy for player B. Using the entries of the given matrix, we obtain the inequalities

$$+ 8s + 3t \leq V,$$
$$6r + 4s + 5t \leq V,$$
$$s + 2t \leq V$$

Dividing by V, we get

$$\frac{r}{V} + 8\frac{s}{V} + 3\frac{t}{V} \leq 1,$$
$$6\frac{r}{V} + 4\frac{s}{V} + 5\frac{t}{V} \leq 1,$$
$$\frac{s}{V} + 2\frac{t}{V} \leq 1$$

subject to the condition

r + s + t = V.

Consequently, we have $\underline{r} + \underline{s} + \underline{t} = \underline{1}$

$V \quad V$

Substitute

$$\frac{r}{V} = x, \quad \frac{s}{V} = y, \quad \frac{t}{V} = w.$$

Then we have the relations

$$x + 8y + 3w \le 1,$$
$$6x + 4y + 5w \le 1,$$
$$y + 2w \le 1.$$

We have to minimise V. i.e, We have to maximise = $\frac{r}{V} + \frac{s}{V} + \frac{t}{V}$. i.e, We have to maximise x + y + w.

Thus, the given game is transformed into the following equivalent LPP:

maximise x + y + w

subject to the restrictions

$$x + 8y + 3w \le 1,$$
$$6x + 4y + 5w \le 1,$$
$$y + 2w \le 1,$$
$$x \ge 0, y \ge 0, w \ge 0.$$

Questions

Explain how a game theory problem can be solved as an LPP.

Transform the game

	Y_1	Y_2
X_1	a_{11}	a_{12}
X_2	a_{21}	a_{22}

into an LPP.

Using simplex method for LPP, solve the following game:

6	2
3	5

Answer: $r = \frac{1}{2}$, $s = \frac{1}{2}$,V=4

Solve the following game with LPP approach:

10	6
4	8

Answer: , , V = 7

****

Goal Programming Formulation

Lesson Outline

Introduction to goal programming

Formulation of goal programming problems

Learning Objectives

After reading this lesson you should be able to

Understand the importance of goal programming

Formulate goal programming problems

Introduction

Generally speaking, the objective of a business organization is to maximize profits and to minimize expenditure, loss and wastage. However, a company may not always attempt at profit maximization only. At times, a necessity may arise to pay attention to other objectives also. We describe some such situations in the sequel.

A manufacturing organization may like to ensure uninterrupted supply of its products even if it means additional expenditure for the procurement of raw materials or personal delivery of goods during truckers' strike, etc. with the objective of assuring the good will of the customers.

A company may be interested in the full utilization of the capacity of the machines and therefore mechanics may be recruited for attending to break downs of the machines even though the occurrence of such break downs may be very rare.

A company, driven by social consciousness, may spend a portion of its profits on the maintenance of trees, parks, public roads, etc. to ensure the safety of the environment, with the objective earning the support of the society.

Another organization may have the objective of establishing brand name by providing high quality products to the consumers and for this purpose it may introduce rigorous measures of quality checks even though it may involve an increased expenditure.

While all the sales persons in a company are formally trained and highly experienced, the management may still pursue a policy to depute them for periodical training in reputed institutes so as to maximize their capability, without minding the extra expenditure incurred for their training.

A travel agency may be interested to ensure customer satisfaction of the highest order and as a consequence it may come forward to operate bus services even to remote places at the normal rates, so as to retain the customers in its fold.

A bank may offer services beyond normal working hours or on holidays even if it means payment of overtime to the staff, in order to adhere to the policy of customer satisfaction on priority basis.

A business organization may accord priority for the welfare of the employees and so a major part of the earnings may be apportioned on employee welfare measures.

A garment designer would like to be always known for the latest fashion and hence may spend more money on fashion design but sell the products at the normal rates, so as to earn the maximum reputation.

A newspaper may be interested in earning the unique distinction of 'Reporter of Remote Rural Areas' and so it may spend more money on journalists and advanced technology for communication.

The above typical instances go to show that the top level management of a business organization may embark upon different goals in addition to profit maximization. Such goals may be necessitated by external events or through internal discussion. At times, one such goal may be in conflict with another goal.

'Goal programming' seeks to deal with the process of decision making in a situation of multiple goals set forth by a business organization. A management may accord equal priority to different goals or sometimes a hierarchy of goals may be prescribed on their importance. One has to strive to achieve the goals in accordance with the priorities specified by the management. Sometimes the goals may be classified as **higher level goals and lower level goals** as perceived by the management and one would be interested in first achieving the higher order goals and afterwards considering lower order goals.

Some of the goals that may be preferred by a business organization are : maximum customer satisfaction, maximum good will of the customers, maximum utilization of the machine capacity, maximum reliability of the products, maximum support of the society, maximum utilization of the work force, maximum welfare of the employees, etc.

Since different goals of an organization are based on different units, the goal programming has a multi-dimensional objective function. This is in contrast with a linear programming problem in which the objective function is uni-dimensional.

Given a goal of an organization, one has to determine the conditions under which there will be under-achievement and over-achievement of the goal. The ideal situation will be the one with neither under-achievement nor over-achievement of the goal.

Formulation of Goal Programming Problems

In the sequel, we consider illustrative situations so as to explain the process of problem formulation in goal programming.

Notations

If there is a single goal, we have the following notations:

Let Du denote the under-achievement of the goal.

Let Do denote the over-achievement of the goal.

If there are two goals, we have the following notations:

Denote the under-achievement and the over-achievement of one goal by Du1 and Do1 respectively.

Denote the under-achievement and the over-achievement of another goal by Du2 and Do2 respectively.

Problem 1

Alpha company is known for the manufacture of tables and chairs. There is a profit of Rs. 200 per table and Rs. 80 per chair. Production of a table requires 5 hours of assembly and 3 hours in finishing. In order to produce a chair, the requirements are 3 hours of assembly and 2 hours of finishing. The company has 105 hours of assembly time and 65 hours of finishing. The company manager is interested to find out the optimal production of tables and chairs so as to have a maximum profit of Rs. 4000. Formulate a goal programming problem for this situation.

Solution

The manager is interested not only in the maximization of profit but he has also fixed a target of Rs. 4000 as profit. Thus, the problem involves a single goal of achieving the specified amount of profit.

Let D_u denote the under achievement of the target profit and let D_o be the over achievement.

The objective in the given situation is to minimize under achievement. Let Z be the objective function. Then the problem is the minimization of $Z = D_u$.

Formulation of the constraints

Let the number of tables to be produced be x and let the number of chairs to be produced be Y.

Profit from x tables = Rs. 200 x

Profit from y chairs = Rs. 80 y

The total profit = Profit from x tables and y chairs

under achievement of the profit target -

over achievement of the profit target

So we have the relationship 200 x + 80 y + D_u - D_o = 4000.

Assembly time

To produce x tables, the requirement of assembly time = 5 x hours. To produce y

chairs, the requirement is 3 y hours. So, the total requirement is $5x + 3y$ hours. But the available time for assembly is 105 hours. Therefore constraint

$$5x + 3y \leq 105$$

must be fulfilled

Finishing time

To produce x tables, the requirement of finishing time = 3 x. To produce y chairs, the requirement is 2 y . So, the total requirement is 3 x+ 2 y. But the availability is 65 hours. Hence we have the restriction

$$3x + 2y \leq 65$$

Non-negativity restrictions

The number of tables and chairs produced, the under achievement of the profit target and the over achievement cannot be negative. Thus we have the restrictions

$$x \geq 0, y \geq 0, D_u \geq 0, D_o \geq 0$$

Statement of the problem

$$\text{Minimize } Z = D_u$$

subject to the constraints

Problem 2

Sweet Bakery Ltd. produces two recipes A and B. Both recipes are made of two food stuffs I and II. Production of one Kg of A requires 7 units of food stuff I and 4 units of food stuff II whereas for producing one Kg of B, 4 units of food stuff I and 3 units of food stuff II are required. The company has 145 units of food stuff I and 90 units of food stuff II. The profit per Kg of A is Rs. 120 while that of B is Rs. 90. The manager wants to earn a maximum profit of Rs. 2700 and to fulfil the demand of 12 Kgs of A. Formulate a goal programming problem for this situation.

Solution

The management has two goals.

To reach a profit of Rs. 2700

Production of 12 Kgs of recipe A.

Let D_{up} denote the under achievement of the profit target.

Let D_{op} denote the over achievement of the profit target.

Let D_{uA} denote the under achievement of the production target for recipe A.

Let D_{oA} denote the over achievement of the production target for recipe A.

The objective in this problem is to minimize the under achievement of the profit target and to minimize the under achievement of the production target for recipe A.

Let Z be the objective function. Then the problem is the minimization of

$$Z = D_{up} + D_{uA}$$

Constraints

Suppose the company has to produce x kgs of recipe A and y kgs of recipe B in order to achieve the two goals.

Condition on profit

Profit from x kgs of A = 120 x

Profit from y kgs of B = 90 y

The total profit = Profit from x kgs of A + Profit from y kgs of B

under achievement of the profit target

– over achievement of the profit target

$$120\,x + 90\,y + D_{up} - D_{op}$$

Thus we have the restriction

$$120\,x + 90\,y + D_{up} - D_{op} = 2700$$

Constraint for food stuff I:

$$7\,x + 4\,y \leq 145$$

Constraint for food stuff II:

$$4\,x + 3\,y \leq 90$$

Non-negativity restrictions

$$x\,,\,y\,,\,D_{up}\,,\,D_{op}\,,\,D_{uA}\,,\,D_{oA} \geq 0$$

Condition on recipe A

The target production of A = optimal production of A

under achievement in production target of A

– over achievement of the production target of A.

Thus we have the condition

$$+\,D_{uA} - D_{oA} = 12$$

Statement of the problem

$$\text{Minimize } Z = D_{up} + D_{uA}$$

subject to the constraints

Questions

Explain the necessity of a goal programming.

Describe some instances of goal programming.

Explain the formulation of a goal programming problem.

Queueing Theory

Lesson Outline

Introduction
Basic concepts of a Queueing System
Numerical Examples of Waiting line Models

Learning Objectives

After reading this lesson you should be able to

Understand the nature and scope of a Queueing System
Waiting line Models and the solutions to the problems of Waiting line Models

Introduction

A business organization may be either production oriented or service oriented or both. With the rapid advancement of science and technology, the service sector has registered a phenomenal growth in the recent years. Customer satisfaction being the hallmark of an efficient service organization, a customer would like to receive the desired service as soon as he enters the organization or at least within a reasonable time of his arrival at the spot. **Queueing theory** is associated with the objective of improving the efficiency of service in any organization engaged in offering a service to the customers. If there are sufficient facilities in the service point to offer the desired service as soon as a customer enters, then there is no necessity to form a queue. Accumulation of customers at a service point leads to the formation of a queue.

The flow of customers from a finite or infinite population towards the service point leads to the formation of a ***queue (waiting line)***. When there is exact match between the service facilities and the arrival of customers, ***waiting time*** is caused either for the service facilities or for the arrival of the customers. A queueing system consists of one or more queues and one or more servers. ***A queueing system*** works with a class of procedures. The incoming customer either waits at the queue sometimes or gets his turn to be served based on the working conditions prevalent in the queueing system. If the server is free at the time of arrival of a customer, then the customer can directly enter into the counter for getting his service. Thus there are two possible things: the system may experience "customer waiting" and /or "idle time of the server".

System Approach to a Queue

A system consists of three components, namely, (i) input, (ii) processes and (iii) output. As regards a queueing system, the input is constituted by the customers who arrive at a service point in anticipation of a service. The process includes the methods of the service offered by the organization, the behavior of the customers and the discipline of the queue. Hence, the following data are required to understand and analyze a queueing system:

3. The input (the pattern of arrival of the customers to the service point)
4. The service mechanism (the pattern of service in the organization)
5. The queue discipline (the principle under which the queue operates) and
6. The behaviour of the customers

The input (the pattern of arrival)

One has to find out the way in which the customers arrive at a service point and join the queue. Customers normally arrive almost in a random way. It is highly difficult to guess the pattern of arrival of the customers. Therefore we have to associate probabilities with the arrival of the customers and hence the probability distribution for ***inter-arrival*** times (the time between two successive arrivals of the customers) has to be found out. We take up a queueing system with the assumption that the customers arrive in a Poisson process. We also assume that the mean arrival rate of the customers is found it to be δ.

The Service Mechanism

The term service mechanism refers to the arrangement of service facility to serve the customers. If there is infinite number of servers, then all the customers are served instantaneously as soon as they arrive and a queue will not be found. If the system consists of a finite number of servers, then the customers are served according to a pre-determined rule by making the server service time a constant or a random variable. Distribution of service time follows '***Exponential distribution***'.

Queueing Discipline

The term queueing discipline refers to a procedure by which the customers are selected from the queue for offering the service. The following disciplines are generally adopted by a queueing system:

First Come First Served – (FCFS)

First In First Out – (FIFO)

Last In First Out – (LIFO)

Selection for service In Random Order (SIRO)

The Behaviour of the Customers

In general, one may observe the following modes of behavior of the customers in a queue:

Normally, the customers arrive one by one into the system. However, there is a possibility for another phenomenon. The term ***Bulk arrival*** refers to the arrival of customers in groups.

Consider the case of several service counters in the organization. Then each service counter will have a queue. When there are several queues, the customers from one queue may switch over to another queue if it is of smaller size. Such a behaviour of the customers is referred to as ***Jockeying.***

Sometimes a customer on arrival may not join a queue after observing that the queue length is very large. This behavior of the customers is called ***Balking*** of the customers.

In certain cases, a customer already present in a queue may leave the queue thinking that the waiting time may be too much. This behaviour of the customers is called ***Reneging.***

Notations

Customers in the system refers to the customers receiving service at the service point and the customers who are waiting to receive the service.

The following notations are used in Queueing theory:

n - No of customers in the system

C - No of servers in the system

$P_n(t)$ – The probability of having n customers in the system at time t

P_n - The steady state probability of having n customers in the system

P_0 - The probability of having zero customer in the system

L_q - Average number of customers waiting in the queue

L_s - Average number of customers waiting in the system (in the queue and in the service points)

W_q - Average waiting time of the customers in the queue

W_s - Average waiting time of the customers in the system (in the queue and in the service points)

10. - Arrival rate of the customers μ - Service rate of the server

ϕ - Utilization factor of the server

11. eff - Effective rate of the arrival of customers M - Poisson distribution

N - Maximum number of customers allowed in the system. It also refers to the size of the calling source of the customers

GD - General discipline for service in the organization like first in first –

served (FIFS), last-in-first served (LIFS), etc.

Traffic Intensity (Or Utilization Factor)

The traffic intensity of a queue is denoted by ϕ. It is defined by the rule ϕ = (Mean arrival time)/(Mean service time) = δ/μ (< 1)

and the unit of traffic intensity is called Erlang.

Types of Waiting Line Models

In general, a queueing model can be classified into six categories using Kendall's notation with six parameters to define a model. The parameters of this notation are as follows.

P- Arrival rate distribution i.e probability law for the arrival /inter – arrival time.

Q - Service rate distribution, i.e probability law according to which the customers are being served

R - Number of Servers (i.e., number of service points)

X - Service discipline

Y - Maximum number of customers allowed in the system

Z - Size of the calling source of the customers

A queuing model with the above parameters is denoted by the symbol (P/Q/R : X/Y/Z)

Model 1 : (M/M/1) : (GD/ ∞/ ∞) Model

This is a waiting line model with the following properties:

11. The arrival rate follows Poisson (M) distribution
12. The service rate follows Poisson distribution (M)
13. The number of servers in the system is 1
14. The service discipline is general discipline (i.e., GD)
15. Maximum number of customers allowed in the system is infinite (∞)
16. he size of the calling source is infinite (∞)

The steady state equations to obtain Pn, the probability of having n customers in the system, and the formulae for obtaining the values for Ls, Lq, Ws and Wq are as follows:

n = 0,1,2,... ∞ where $\phi = \delta/\mu < 1$

Ls – Average number of customers waiting in the system (i.e., waiting in the queue and the service station)

Pn= ϕ_n (1-ϕ)

Ls= ϕ/(1- "ϕ")

Lq= Ls- δ/μ-

2. ϕ/(1- "ϕ") - ϕ

3. (" ϕ" - (1-"ϕ") "ϕ")/(1-"ϕ")

4. ϕ_2/(1-"ϕ")

Average waiting time of customers in the system (in the queue and in the service point) = Ws

3. Ls/δ = ϕ/((1 - "ϕ")"δ")

4. ϕ/((1 - "ϕ")) x 1/$\mu\phi$ (since $\delta = \mu\phi$)

5. 1/("μ" -"$\mu\phi$")

6. 1/("μ" -"δ")

Wq = Average waiting time of customers in the queue

4. Lq/δ = 1/δ x ϕ_2/(1-"ϕ")

5. 1/$\mu\phi$ x ϕ_2/(1-"ϕ") (since $\delta = \mu\phi$)

6. ϕ/("μ" -"$\mu\phi$")

7. ϕ/("μ" -"δ")

Summary of Formulae

5. $\phi = \delta/\mu$
6. $Pn = \phi^n (1-\phi)$
7. $P0 = 1 - \phi$
8. Ls=ϕ/(1- "ϕ")
5. Lq = ϕ_2/(1 - "ϕ")
6. Ws = 1/("μ" -"δ")
7. Wq = ϕ/("μ" -"δ")

Example 1

The arrival rate of customers at a petrol bunk follows a Poisson distribution with a mean of 27 per hour. The petrol bunk has only one unit of service. The service rate at the petrol bunk also follows Poisson distribution with mean of 36 per hour. Determine the following:

What is the probability of having zero customer in the system ?

What is the probability of having 6 customers in the system ?

What is the probability of having 10 customers in the system ?

The values of Ls, Lq, Ws and Wq

Solution

Given that the arrival rate follows Poisson distribution with mean = 27

With our notations, $\delta = 27$ per hour

Given that the service rate follows Poisson distribution with mean = 36

Hence $\mu = 36$ per hour

Consequently, the utilization factor $\phi = \delta/\mu = 27/36 = 3/4 = 0.75$

The probability of having zero customer in the system is

P0 = ϕ_0 (1- ϕ)

7. 1- ϕ
8. 1- 0.75
9. 0.25

The probability of having 6 customers in the system is

P6 =ϕ6 (1- ϕ)

8.(0.75)6 (1- 0.75)

9.0.0178 x 0.25

10. 0.0445

The probability of having 10 customers in the system is

P10 = ϕ10 (1- ϕ)

10. (0.75)10 (1- 0.75)

11. 0.0563 x 0.25 = 0.0141

Ls = ϕ/(1- "ϕ")= 0.75/(1- 0.75) = 0.75/0.25 = **3 customers**

Lq = ϕ2/(1-"ϕ") = (0.75)^2/(1-0.75) = 0.5625/0.25 =**2.25 customers**

Ws = 1/("μ" -"δ") = 1/(36 -27) = 1/9 =**0.011 hour**

Wq = ϕ/("μ" -"δ") = 0.75/(36-27) = 0.75/(9)= **0.0833 hour**

Note: In the determination of Ls and ¬ Lq, we are not rounding the values to the nearest integer.

Example 2

At one-man book binding centre, customers arrive according to Poisson distribution with mean arrival rate of 4 per hour and the book binding time is exponentially distributed with an average of 12 minutes. Find out the following:

The average number of customers in the book binding centre and the average number of customers waiting for book binding.

The percentage of time arrival can walk in straight without having to wait.

The percentage of customers who have to wait before getting into the book binder's table

Solution

Given mean arrival of customers $\delta = 4/60 = 1/15$

Mean time for server $\mu = 1/12$

Hence $\phi = \delta / \mu = [1/15] \times 12 = 12 / 15 = 0.8$

The average number of customers in the system (Customers in the queue and in the book binding centre)

$$Ls = \phi / 1 - \phi = 0.8 / 1 - 0.8$$

$= 0.8 / 0.2$

$= 4$

$= 4$ Customers

The percentage of time arrival can walk straight into book binder's table without waiting is

service utilization $= \phi\ \%$

$\delta/\mu\%$

$0.8 \times$

$100 = 80$

The percentage of customers who have to wait before getting into the book binder's table $= (1-\phi)\%$

$(1-0.8)\%$ $= 0.2 \times 100$

$= 20\%$

Example 3

A person repairing wrist watches observes that the time spent on the wrist watches has an exponential distribution with mean 20 minutes. If the wrist watches are repaired in the order in which they come in and their arrival is Poisson with an average rate of 15 for 8-hour day, what is the repairman's expected idle time each day? On an average, how many jobs are ahead of a wrist watch just brought in?

Solution

The arrival rate $\delta = 15/8 \times 60 = 1/32$ units/minute

The service rate $\mu = 1/20$ units/minute

$\phi = \delta/\mu$

1/32 / 1/20

20/32 = 5/8

The number of jobs ahead of the wrist watch just brought in = the average number of jobs in the system

Ls =ϕ/1-ϕ

□(5/8) / {1-(5/8)}

□(5/8) /(3/8)

□5/3.

The number of hours for which the repairman remains busy in an 8-hour day = 8 x δ/μ = 8x (5/8) =5 hours.

Hence the time for which repairman remains idle in an 8-hour day = 8-5= 3 hours.

Example 4

Customers are arriving at a service centre at the rate of 70 per hour. The average service time for a customer is 45 seconds. The arrival rate and service rate follow Poisson distribution. There is a complaint that the customers wait for a long duration. The proprietor of the centre is ready to consider the installation of one more service point to reduce the average time to 35 seconds if the idle time of the service point is less than 9% and the average queue length at the service point is more than 8 customers. Determine whether the installation of the second service point is worth from the point of view of the proprietor.

Solution

Arrival rate of the customers at the service centre δ =70 per hour

Average service time for a customer = 45 Seconds

$$\text{Service rate} \quad \mu = \frac{1 \text{ hour}}{45 \text{ seconds}}$$

= 3600/45

= 80

= 80 customers per hour

Utilization factor $\phi = \delta/\mu$

$= 70/80$

$= 0.875$

(a) Waiting no. of customers in the queue is Lq

$$Lq = \phi_2/((1 - \text{"}\phi\text{"})) = \frac{(0.875)2}{1-0.875} = \frac{0.7656}{0.125} = 6.125$$

= 6 Vehicles

Desired service time for a customer (after addition another service point) =35 seconds
The new service rate after installation of an additional service point = 1 hour/35 Seconds= 3600/35

=102.68 Customers / hour

Utilization factor $\phi = \delta/\mu = 70 / 102.86$

$= 0.681$

Percentage of idle time of the service point = $(1-\phi)\%$

$(1-0.681)\%$

0.319%

31.9

32%

This idle time is not less than 9% which is desired by the proprietor. Hence, the installation of the second service point is not justified since the average waiting number of customers in the queue is more than 8 but the idle time is not less than 32%.

Model 2 : (M/M/C) : (GD/ ∞/ ∞) Model

The following assumptions are made for this model:

The arrival rate follows Poisson distribution

The service rate follows Poisson distribution

The number of servers is C

The service discipline is general discipline

The maximum number of customers allowed in the system is infinite

With these assumptions, the steady state equation for the probability of having n customers in the system is given by

$$Pn = \frac{\phi^{n} Po}{n!} \quad , o \leq n \leq C$$

$$= \frac{\phi n \ Po}{\square n\text{-}C \ C!} \quad \text{for } n > C \text{ where } \phi / C < 1$$

where $[\delta / \mu C] < 1$ as $\phi = \delta / \mu$

$$P0 = \{[\sum_{n=0}^{C-1} \phi_n / n!] + \phi C / (C! [1 - \phi / C])\}\text{-}1$$

where $C! = 1 \times 2 \times 3 \times \ldots\ldots\ldots\ldots\ldots$ upto C

$$Lq = [\phi \, C+1 / [C\text{-}1! \, (C - \phi)]] \times P0$$

$$= (C\phi PC) / (C\text{-}\phi)2$$

$$Ls = Lq + \phi \quad \text{and } Ws = Wq + 1 / \mu$$

$$Wq = Lq / \delta$$

Particular cases

Under special conditions $Po = 1 - \phi$ and $Lq = \phi_{c-1} / C\,2$ where $\phi < 1$ and $Po = (C\text{-}\phi)\,(C - 1)! / C\,C$

and $Lq = \phi / (C\text{-}\phi)$, where $\phi / C < 1$

Example 1

At a Toll Gate, vehicles arrive at the rate of 24 per hour and the arrival rate follows Poisson distribution. The time to collect a toll and permitting the vehicle to pass follows exponential distribution and the passing rate is 18 vehicles per hour. There are 4 passing counters. Determine the following:

Po and P3

Lq, Ls, Wq and Ws

Solution

The arrival rate δ= 24 per hour.

The passing rate μ = 18 Per hour.

No. of passing counters C=4.

2. =δ/μ

24/18

=1.33

C-1

3. P0 ={[Σ ϕn/n!] + ϕC / (C! [1 - ϕ/C])}-1 n

= 0

3

={[Σ (1.33)n/n!]+ (1.33)4 /(4! [1 - (1.33)/ 4])}-1

n = 0

={ (1.33)0 / = 0! + (1.33)1 / 1! + (1.33)2 / 2! +

(1.33)3 / 3! + (1.33)4 / 24! [1 - (1.33)/ 4] }-1

=[1 + 1.33 + 0.88 + 0.39 + 3.129/16.62] -1

=[3.60 + 0.19]-1

=[3.79]-1

= 0.264

We know Pn = (ϕn / n!) Po for $0 \leq n \leq C$

P3 = (ϕ3 / 3!) Po since $0 \leq 3 \leq 4$

= (1.33)3 / 6] x 0.264

= 2.353 x 0.044

= 0.1035

(ii) Lq = $\dfrac{\phi_{c+1} X P0}{(C-1)!\,(C-\phi)2}$

$$= \frac{(1.33)5 \times 0.264}{3! \times (4 - 1.33)2}$$

$$= \frac{(4.1616) \times 0.264}{6 \times (2.77)2}$$

$$= \frac{(4.1616) \times 0.264}{46.0374}$$

= 1.099 / 46.0374

= 0.0239

= 0.0239 Vehicles

Ls = Lq + ϕ = 0.0239 + 1.33

= 1.3539 Vehicles

Wq = Lq / δ = 0.0239 /24

= 0.000996 hrs

Ws = Wq + 1 /μ = 0.000996 + 1/18

3. 0.000996 + 0.055555

4. 0.056551 hours.

Example 2

In a bank, there are two cashiers in the cash counters. The service time for each customer is exponential with mean 4 minutes and the arrival rate of the customers is 10 per hour and the arrival of the customers follows Poisson distribution. Determine the following:

4. The probability of having to wait for service
5. The expected percentage of idle time for each cashier
6. Whenever a customer has to wait, how much time he is expected to wait in the Bank?

Solution

C-1

P0 =$\{[\Sigma\phi n/n!] + \phi C / (C! [1 - \phi/C])\}$-1 n

= 0

Where $\phi = \delta / \mu$ given arrival rate = 10 per hour =

10 / 60 = 1 / 6 per minute

Service rate = 4 minutes

$\mu = 1 / 4$ person per minute

Hence $\phi = \delta / \mu$ $= (1 / 6) \times 4 = 2 / 3$

$= 0.67$

1

P0 $=\{[\Sigma\ \phi n/n!]+(0.67)2 / (2! [1 - 0.67/2])\}-1$ n

$= 0$

$=[1 + (\phi / 1!)] + 0.4489 / (2 - 0.67)]-1$

$=[1 + 0.67 + 0.4489 / (1.33)]-1$

$=[1 + 0.67 + 0.34]-1$

$=[2.01]-1$

$1 / 2$

2 The Probability of having to wait for the service is P (w > 0)

$$\frac{\phi C\ XP0}{C!\ [1 - \phi/C]}$$

$$\frac{0.67\ 2X\ (1 / 2)}{2!\ [1 - 0.67 /2]}$$

$0.4489 / 2.66$

0.168

3 The probability of idle time for each cashier is

1- P (w > 0)

1-1/3

2/3

Percentage of time the service remains idle = 67% approximately

iii. The expected length of waiting time (w/w>0)

$=1/(C\mu-\delta)$

$=1/ [(1/2)-(1/6)]$

=3 minutes

Questions

Explain the basic concepts of Queuing theory.

The arrival rate of customers at a booking counter in a railway station follows Poisson distribution with a mean of 45 per hour. The service rate of the booking clerk also follows Poisson distribution with a mean of 60 per hour.

i. What is the probability of having 0 customer in the system?

What is the probability of having 5 customers in the system?

What is the probability of having 10 customers in the system?

Determine Ls, Lq,Ws and Wq

Answer

0.25

ii.0.0593

iii.0.0141

Ls=3 customers, Lq=2.25 customers, Ws =0.067 hour and Wq = 0.05 hour.

The arrivals and services in a service centre follow Poisson Distribution. The arrival rate of the customers is 8 per hour. The service rate is 10 customers per hour. Find out the following:

i. The average number of the customers waiting for service.

The average time a customer has to wait in the queue.

The average time a customer has to be in the system.

Answer

i. 3.2 customers

0.4 hours

iii.0.3 hours

www.ingramcontent.com/pod-product-compliance
Lightning Source LLC
LaVergne TN
LVHW060027210926
848481LV00013B/3807
9798215113431